Q^{the}uantum Doctor

the Quantum Doctor

A Quantum Physicist Explains
the Healing Power of Integrative Medicine

Amit Goswami, PhD

HAMPTON ROADS

Hampton Roads Publishing Company, Inc.
Charlottesville, VA 22906
www.hrpub.com

Library of Congress Cataloging-in-Publication Data available on request

ISBN : 978-1-57174-655-9

10 9 8 7 6 5 4 3
TS
Printed on acid-free paper in United States of America

*This book is dedicated to
the healing of the planet Earth*

Table of Contents

Part 3. Mind-Body Medicine

Part 4. The Healing Path to Supramental Intelligence

Foreword

When I was a young intern in Boston—this was almost forty years ago—an older couple was admitted to the hospital together. The wife was in the final stages of cancer; the husband was suffering from a much milder respiratory distress. I saw them every day and was struck by their closeness. It would go very hard on the husband when his wife was gone. The days passed. She quickly declined; the husband's condition, although not life-threatening, stubbornly resisted treatment. Finally I had to break the news to him that his wife had passed during the night. He nodded, and one could only guess at the pain he concealed.

I expected to say goodbye to him in a few days, but something remarkable happened. His condition grew worse, then critical, and within three days he was dead. Baffled, I went to my supervisor, who had decades of experience in hospital care. He said, "Don't you understand? He was ready to die. Anyway, he had to go after her. A gentleman always waits for a lady."

I still hear him saying those words, but now I am not a baffled young doctor. The way that people relate to their bodies, which includes the bewildering territory we call illness and wellness, remains mysterious. In this remarkable book, Amit Goswami penetrates the mystery, and he does it better than anyone before or since. We should all pay attention, because the mind-body connection holds enormous promise. Placebo, for example, has become an everyday word, but the power of the placebo effect has yet to be harnessed.

In one study a group of patients suffering from chronic severe nausea were given a drug that they were told was an anti-nausea agent. Normally in a placebo trial half the patients would be randomly selected to receive the drug while the other half received a dummy pill. As expected, more than 30% of the subjects who took the dummy pill reported that their nausea was alleviated. Only this was a placebo trial with a difference: instead of the typical sugar pill, they had been given an emetic, a drug that promotes nausea. So powerful was their belief, however, that it somehow reduced their symptoms—or to be more accurate, their minds reduced nausea in the face of a drug that should have made them vomit.

Professor Goswami is bold enough to aim directly at the source of placebo, and all other forms of mind-body treatment, when he asserts that all of reality, including the human body, is based in consciousness. In this, he has joined ancient wisdom with cutting-edge physics.

Two thousand years ago the philosophy of Vedanta declared that material existence is an illusion, a shared dream, from which it is possible to awaken, and when we do, we realize that behind the illusion was pure consciousness. Such a view had little bearing on Western thought until the great quantum pioneers arrived at the beginning of the twentieth century. Their names are celebrated today—Albert Einstein, Erwin Schrödinger, Wolfgang Pauli, Werner Heisenberg–but what is much less well known is that almost all became mystics. Having discovered that the solid material world was based on invisible energy fields, and that those fields emerge from a place outside space and time, the quantum pioneers began to alert the public that the physical world was shifting under our feet like quicksand. Niels Bohr declared, "Everything we call real is made of things that cannot be regarded as real."

Heisenberg said in his Nobel Prize speech of 1932 that the atom has "no physical properties at all." Einstein posited that everything in the universe was happening in the mind of God.

For those of us who are outside physics, it's incredibly helpful that someone as open-minded as Amit Goswami should step forward to take this revolution seriously, for if the atom isn't physical, neither is the universe, and neither is the human body. Mainstream physics has largely ignored this startling idea, preferring to follow the advice of a well-known researcher, "Shut up and calculate." Goswami stands out as a speculative thinker who won't shut up. In this new book he expands on the simplest and yet most profound hypothesis: if the body isn't a thing at all, how far can medicine go in treating it as something else?

But what would that something else be? That's where having an expert in quantum physics proves invaluable. When you delve into the microscopic world where matter disappears, at the horizon of space and time when both return into the vacuum state, the void that is actually the womb of creation, reality isn't about to vanish. Quite the opposite—as the ancient Vedic *rishis* taught, creation is richest at its source, because it is here that consciousness consists of infinite possibilities. If the mind could manipulate those possibilities, we humans would find ourselves to be co-creators of the physical world that is projected all around us.

More specifically, we could create events in the body. Instead of being the victims of illness and disorder, we could return to a state of healthy balance and vitality. There is no doubt, forty years after the outset of the mind-body revolution in medicine, that the physical tools exist for connecting mind and body. A mountain of research into messenger molecules reveals that the brain's most minute activity is translated into chemicals that carry the same activity to every cell in the body. Thanks to hundreds of thousands of receptor sites on the outer membrane of each cell, we have no doubt that the qualities once ascribed only to the brain—including intelligence and conscious awareness—are shared with the rest of the body.

Why, then, do we get sick when obviously our brains, acting as agents of the mind, want to be well? Hundreds of answers exist, most of them plausible. Perhaps we are suffering from toxic emotions. Perhaps we have a genetic predisposition that the mind cannot override. The problem is that there is no coherent theory that can serve as a foundation to explain how the mind heals the body, or how it fails to. The mind-body field is hit and miss when it comes to achieving results. In fact, the most highly touted approaches of alternative medicine rarely perform better than the placebo effect.

The Quantum Doctor fills the need for an overarching theory of mind and body with courage and intelligence, along with a deep background in both ancient Indian thought and modern physics. Prof. Goswami immediately cuts through the seductiveness of materialism.

Mainstream medicine, he rightly points out, is consistent in its philosophy, which holds that the body is a material object existing in the physical world. On the basis of materialism, Western scientific medicine has been triumphantly successful at conquering many diseases. Germs can be killed with drugs; damaged hearts can be repaired with bypass surgery. But behind this success lurks the unsolved mystery of the mind, which materialism cannot touch.

Goswami holds that a subtler kind of materialism, such as the use of herbs or the manipulation of Chi, the Chinese term for the life force, would lead us in the wrong direction. Here I wholly support him. Whether mainstream medicine and its chief ally, big pharma, like it or not, the human body is controlled by the mind. In at least four thousand cases of spontaneous remission from cancer, a patient's desire to get well led to a cure, sometimes overnight, without the intervention of drugs and surgery. Since oncologists continue to ignore these remarkable cures, by and large, this book can fill a vacuum that much needs filling.

The reason a physicist must step in is this: when a thought activates a molecule in the brain, it is actually performing a quantum operation. A molecule (for example, dopamine or serotonin) emerges from nothing, and the combination of millions of such molecules becomes the physical counterpart of thoughts, intentions, desires, wishes, hopes, and dreams. What I've stated is literal fact. If you take a patient suffering from obsessive compulsive disorder (OCD), the standard treatment is a drug like Prozac. On a brain scan, one sees the area of the brain that is unbalanced in OCD begin to behave more normally. Yet at the same time, if a patient doesn't take a drug but goes to a psychiatrist instead, talking about the problem and going into the personal roots of OCS brings relief. And in those cases, a brain scan reveals that the same area of the brain has returned to normal.

At the very least, the brain runs on dual control, responding to drugs as well as to immaterial things like words. But we should have admitted this long ago in medicine. Materialism holds that depression, for example, is the result of a chemical imbalance in the brain, and antidepressants are based on that assumption that chemicals are needed to correct chemicals (never mind that the most up-to-date research shows that the brains of depressed people are not chemically imbalanced in the way that theory suggested, nor do popular antidepressants correct such imbalances). The fact is that I can make a person depressed simply by speaking words. I can tell him that he's lost his job or that all the money in his bank account is gone. This indisputable fact leads us to the place that *The Quantum Doctor* elucidates, the junction point where the material world is subject to immaterial forces.

I will leave to the reader the exciting discovery that consciousness holds the switch to wellness. It's to Prof. Goswami's credit that he has stripped away the intellectual haze that surrounds exotic concepts like Prana, vital energy, and the substrata of Ayurveda. What emerges is the very thing we need: a coherent philosophy that demolishes the wall between physics and metaphysics. In a consciousness-based universe such a wall never existed in the first place.

About the time that I was a young resident in Boston, at another hospital another young resident was tending a dying man. Physicians become accustomed to awaiting the timing of life and death, which is ruled only by itself. This doctor walked into the room just as his patient expired. At that instant, as if seeing a glimmer out of the corner of his eye, he witnessed something faint, like a thermal ripple glistening off the highway in summer, slip out of the dying man.

"I was dumbstruck, but I know what I saw," he related afterward. "It was his soul. I saw a soul leaving someone's body."

He never forgot the experience, which had a telling effect on his decision to become a psychiatrist. At that time, four decades ago, psychiatry was as close as anyone could come to explaining the far reaches of the psyche. Now Amit Goswami has taken us much farther. The term he favors, integral medicine, has caught on fairly widely. But in this book the real victory is to entirely erase the dualism of mind and body. That is the holy grail of medicine in general, to find one basis that will explain why drugs and surgery work, why herbs and hands-on healing work, why homeopathy and energy medicine work—indeed, why so many disparate approaches can lead to a cure.

Taking us through the various levels of energy that form the hierarchy of creation, which can also be mapped as the "subtle bodies" underlying the physical body, *The Quantum Doctor* eliminates the need for any medical modality to claim that it is the only true way. Such arrogance can be dispensed with. We can also throw out the hostile opposition that mainstream medicine has often shown to integral medicine, which was largely based on ignorance. What this book offers is complete clarity. All of us desperately need it, and I look forward with much greater hope that one day soon, medical students will be required to read *The Quantum Doctor* before they graduate. This is information that could change the world of medicine, if not the world as a whole.

Deepak Chopra, MD
Spring 2011

Preface

Let me admit at the outset that I write this book as a theoretician, a quantum physicist who sees medicine as a ripe and timely area of application for the new paradigm of science based on the primacy of consciousness. As the reader is probably aware, this new science has a spectacular ability to integrate many disparate fields of human endeavor, even science and spirituality.

If any field needs integration, it is medicine. If any field needs an integrative paradigm that can make sense out of all the different models of healing, it is medicine. The weaknesses of the conventional medical model have been clear for some time. Its procedures are too invasive and have too many harmful side effects. There is no conventional medical model for the treatment of most chronic and degenerative diseases (germ theory and genetic predisposition are not adequate explanations for most conditions in this category). Last, but not least, conventional medicine is expensive.

In contrast, there are so many different alternative medicine models based on so many different philosophies! I will mention three. Mind-body medicine talks about the mind as slayer and healer. Chinese medicine posits disease and healing as the problem and solution, respectively, of the movement of a mysterious energy called *chi.* Indian medicine, Ayurveda, calls disease the imbalances of mysterious attributes that we have—the *doshas*—and sees the solution in the correction of these imbalances.

What criteria do we use in choosing from these different styles of medicine? At least conventional medicine is based on one philosophy—material realism (everything is based on matter which is the only reality)—so conventional medical people can consult with one another without philosophical conundrums. In alternative medicine, that luxury is lacking.

There is some attempt to define a "holistic metaphysics" as a basis for alternative medicine; it is based on the idea that the whole is greater than the parts. But this philosophy suffers from a fundamental materialist prejudice that mind and *chi,* although not reducible to the parts, are nonetheless ultimately material in origin; they are conceived as emergent causal properties of matter not reducible to the components. Because of this materialist prejudice, this kind of holism has been neither popular nor successful.

And yet if you try to understand alternative medicine with materialist metaphysics, you get paradoxes. Additionally, there is much anomalous data, the most famous being the data on spontaneous healing; the overnight cure of cancer without any medicine is an example of spontaneous healing. This, too, the materialist paradigm of medicine cannot explain. A paradigm shift is needed.

Fortunately, help is coming from an unexpected direction. For some time, a new paradigm of physics, quantum physics, has been pointing out the conceptual incompleteness of material realism—the metaphysics favored by conventional medicine. The physician Andrew Weil once entitled a chapter in a book *What Doctors Can Learn from Physicists.* What Weil refers to is the major paradigm shift that physics has been undergoing for quite some time. Recently, this paradigm shift has taken a new turn, and it is becoming obvious that not only is the new physics important for traditional physics and chemistry, but its message must be incorporated in biological sciences as well.

This opens the question: Can the new physics integrate the disparate models of conventional and alternative medicine? In this book, I show that the answer is yes.

In my earlier books, thinking as a quantum physicist, I developed a new way of doing science that I call *science within consciousness*. It is a science based on the primacy of consciousness; consciousness is posited as the ground of all being; and all the quantum paradoxes that you hear about are resolved when quantum physics is formulated within this metaphysics. Meanwhile, other researchers have been busy establishing the need for extraphysical realms of experience. Roger Penrose has shown that computers cannot simulate the one characteristic that defines mind—meaning. So mind must be extraphysical, must be independent of the brain. Rupert Sheldrake has posited extraphysical morphogenetic fields to account for morphogenesis in biology. I myself have shown that a proper study of creativity data clearly indicates the existence of yet another extraphysical body called supramental intellect. Psychologist Carl Jung theorized this domain to be that of our intuition.

In this book I show that when medicine is based on the primacy of consciousness, taking into account all these "bodies" of consciousness (morphogenetic fields, mind, and supramental, besides the physical), both conventional and alternative medicine can be formulated in their appropriate niche, and more. When we use quantum physics as the basis of our formulation of medicine, the old argument of "dualism" posed by conventional medicine against the validity of positing nonphysical mental and other bodies in our theories is no longer valid.

The new paradigm of medicine, which I call Integral Medicine, shows clearly how mind-body healing works, how the medical systems of China and India work, how homeopathy works. Integral Medicine also gives us broad hints as to how to use all these healing practices, including conventional medicine, together, as needed.

How is Integral Medicine as developed here different from the integrative medicine that many other authors explore? You may think of Integral Medicine as integrative medicine because the objective of both is the same. However, existing models of integrative medicine use what is called a "systems theory" approach to join together disparate models. Integral Medicine does the joining based on an integration of the underlying metaphysics of all the models of medicine including conventional allopathy. This is a very new approach. This is a very new accomplishment. This can be a legitimate basis for a paradigm shift in medicine.

Even within the profession of medicine, some physicians, notable among them Andrew Weil, Deepak Chopra (who gave us the wonderful phrase "quantum healing"), and Larry Dossey, are already exploring quantum aspects of healing. This book integrates all this early work as well.

I discuss theory, I discuss new data, I explain concepts, methods, and techniques of alternative medicine, and I explain spontaneous healing. I even discuss the spiritual component of healing. I discuss the problem of death and dying from this new perspective, and I discuss immortality, or what I call the ageless body. But most of all I give the reader a sense of what disease *means,* what healing means, and how to be intelligent about disease and healing. First and foremost, the book is about helping you, the reader, to make sense out of the disparate literature of medicine—conventional and alternative—and to find the path to positive health.

I thank Uma, my wife, for being the inspiration for this book and for contributing to it in many intangible and tangible ways. I thank my editor, Richard Leviton, for asking me to write the book and for giving me excellent feedback, and the staff at Hampton Roads for doing a fine job of production. Finally, I thank all the healing professionals who have always been a constant source of encouragement and inspiration for me.

PART 1

Introducing the Quantum Doctor

1

Never Fear, the Quantum Doctor Is Here

What is a quantum doctor? A quantum doctor is a practitioner of medicine who knows the fallacies of the Newtonian classical physics-based deterministic worldview that was discarded in physics many decades ago. A quantum doctor is grounded in the worldview of the new physics, also called quantum physics. And there is more. Quantum doctors bring the message of quantum physics alive in their practice of medicine.

You may wonder: What difference does a worldview make in the practice of medicine? In contrast to the classical physics worldview in which the world is seen as a mechanical, determined machine, we cannot even make sense of quantum physics unless we ground it in the primacy of consciousness: Consciousness comes first; it is the ground of all being. Everything else, including matter, is a possibility of consciousness. And consciousness chooses out of these possibilities all the events we experience.

Now do you *see*? Physicians of the old ilk of classical physics aficionados practice *machine medicine,* designed for machines (that is the picture of the patient in the classical worldview) and by machines (the physicians who are self-avowed machines). And make no mistake about it, the medicine that the patient gets, allopathic medicine, is also of a mechanical nature, with chemical drugs, mechanical surgery or organ transplant, and energy radiation. A quantum doctor, on the other hand, practices conscious medicine designed for people, not machines. What conscious medicine prescribes includes the mechanical but extends also to the domains of vitality and meaning, even love. And most important, as practitioners of conscious medicine the quantum doctors bring consciousness to their practice.

Admittedly, the *quantum doctor* is right now only an idea developed in this book, an idea you are probably reading about for the first time. But if the idea is here, and as I will show, it is an idea with much integrative power, can the manifestation of the idea be far behind? In fact, a partial manifestation of the idea is age-old and continues to the present era.

I am talking about practitioners of what today is called alternative or complementary medicine, including the age-old systems of acupuncture and traditional Chinese medicine, Ayurveda (developed in India), spiritual healing, the more recent homeopathy, and the very recent mind-body medicine. Alternative medicine practitioners go partway toward being quantum doctors. Their medical systems are designed for conscious beings, and they do have more dimensions than the mechanical. Unfortunately, alternative medicine practitioners suffer from considerable amounts of worldview confusion (discussed later).

Although our culture promotes the "great" advancements of machine medicine all the time, still many people are disillusioned with it. Partly, it is because we all miss the conscious human touch that we expect from a healer. Partly, it is because, its "miracles" notwithstanding, allopathic medicine doesn't work well for the bulk of our day-to-day medical problems—the chronic ailments, for example. And it is because machine medicine and mechanical procedures are very expensive.

So although the machinists in medicine are openly scornful of alternative practices, alternative medicine is gaining in popularity.

Unfortunately, this only aggravates the reaction of the conventional allopathic practitioners. Before, allopaths could afford benign neglect. But now, as their bread and butter is threatened, for many of them it is all-out war against alternative medicine. Alternative medicine is voodoo medicine, they declare.

If the world is machine, mind is machine, and even the soul is machine, as some observers contend, then how can anything but machine medicine have any validity?

Alternative medicine practitioners also strike back. Let's note just two of their criticisms. Allopathic drugs have harmful side effects, they point out. Why should we unnecessarily poison the body? Allopathic procedures such as vaccinations administered when we are children weaken the immune system so much that we become more vulnerable to disease later in life. Why should we accept such procedures without questioning them?

We all are interested in health and healing, in our physical well-being. We all search for it when we don't have it. But with the sharp division of medicine into two camps—conventional and alternative—it is increasingly difficult to choose the proper healing method when we need it. What criteria should we use for such a choice? Is a combination of healing techniques better than any one technique? What should we do to maintain our health, to prevent disease in the first place? Can we heal ourselves without any physical or chemical instruments of healing?

The answers to such questions depend on whom you ask. Are at least some of the stories of spontaneous healing true? Some experts answer yes. But is spontaneous healing accessible to all of us? Experts from some medical traditions nod yes, while others stubbornly shake their heads no. When we are middle-aged or old, are we to feel fortunate if we suffer from only a few chronic diseases, without any major life-threatening illness? Should we accept stress and lack of vitality as the price we must pay for modern living? Maybe so, some experts say. Why is economics such an important consideration for health and healing? We are sorry, say the experts. Is medicine only about pathology? Can we not strive for positive health where vitality and well-being reign supreme? We don't know, say the experts.

Truth is, we cannot begin to answer such questions with much credibility without developing an integral paradigm that embraces all medical systems that have adequate clinical data to support their efficacy. We must end the current confusion of paradigms that pervades medicine.

Never fear, the quantum doctor is here. The quantum doctor, like his or her worldview, is integrative. In this book, I show that when medicine is formulated within the integral metaphysics of the primacy of consciousness, conventional (allopathic) medicine and alternative medicine (including those already listed) can be reconciled. Not only that, their different domains of applicability, even their interrelationships, are clearly understood.

Some effort toward integration has already begun (Ballentine 1999; Grossinger 2000), but without the benefit of an integral philosophy, the results are not convincing. Medicine within the quantum worldview of primacy of consciousness, which gives satisfying answers to all the questions posed in the previous paragraphs, can end the paradigm wars in medicine because it defines a new and consistent paradigm for all of medicine within an integral philosophy.

Is this good news too good to be true? Don't worry. Medicine within quantum consciousness is an offshoot of a general upheaval, a genuine paradigm shift, far greater in scope than even the Copernican revolution, that is taking place today in all of the sciences, including physics, chemistry, biology, and psychology.

Definitions

Some more detailed definitions are in order, although it is likely that the reader is already familiar with them.

Conventional medicine or *allopathy* is based on the premise that disease is due either to external toxic agents such as germs (bacteria and viruses) or to the mechanical malfunctioning of an internal organ of the physical body. In allopathy, cure is effected mainly by treating the symptoms of the disease until they disappear, via drugs, surgery, and (in the case of cancer) energy radiation. Exotic new techniques such as gene therapy or nanotechnology, the premise of which is to correct the mechanical disorder at the molecular level, remain only science fiction.

In contrast, in *mind-body medicine,* the premise is that disease is due to a mental problem, for example, mental stress. The cure is to correct the problem with the mind so that it then will correct the physiology.

In the view of *acupuncture,* disease arises because of imbalances in the patterns of energy *(chi)* flow in the body. Cure consists of correcting these imbalances using skin puncture with tiny needles at appropriate points of the body. The energy referred to in acupuncture is "subtle energy," not to be confused with the usual manifestations of energy, which are "gross."

Acupuncture is the most well-known example of *traditional Chinese medicine,* a system that, in addition to acupuncture, uses special herbs to correct the imbalances in the movements of this subtle energy.

In *homeopathy,* the basic idea is "like cures like" in contrast to cure by "other" (the drug found by trial and error) in allopathy. The same substance that produces gross clinical symptoms in a healthy person when applied in a much-diluted and potentized concentration produces an alleviation of the same symptoms in an unwell person, which is why homeopathy says, "like cures like." But the cure is made mysterious by often (successfully) applying the medicinal agent in such dilutions as one part in 10^{30} or even more diluted.

Ayurveda is traditional Indian medicine. Thanks to the work of such luminaries as the physician Deepak Chopra (2000), Ayurvedic concepts such as *doshas* have become the subject of parlor games. Who are you, a *vata* person, a *pitta* person, or a *kapha* person? *Vata, pitta,* and *kapha* are the Sanskrit names of the three *doshas,* imbalances of bodily structure and movement that we all have in varying degrees. A unique dominance of one *dosha* or another or sometimes a combination of *doshas* characterizes each one of us. In fact, we all have a base level amount of each of the *doshas.* We remain healthy when our *doshas* remain near our individual base-level amounts. Disease happens when deviations occur, taking the body away from the base levels. Returning the body to the base level of the *doshas,* using herbs, massages, and cleansing techniques, effects cure.

Spiritual healing is the idea of invoking the "higher" power of the Spirit through prayer and other such rituals to heal (Holmes 1938). Shamanic healing, prayer healing, Christian Science, faith healing, and intuitive healing fall in this category.

You can see the difficulty many practitioners of conventional medicine have with the various alternative medicine practices as defined here. Mind-body medicine seems to predicate that a mental thought, presumably a brain phenomenon involving a minute amount of energy, can cause disease or healing, which according to conventional medicine requires the emission of neurochemicals and other physiological processes that involve large amounts of energy. "Absurd!" may be the allopathic practitioner's reaction. Chinese medicine talks about subtle energy, but what is this subtle energy? Why can't we find it in the body or the passageways (called meridians) through which it moves? Because they don't exist, the conventional medicine person declares in exasperation.

Similarly, if you are scientifically minded and want to understand the relationship of conventional medicine and *dosha* medicine, you will be disappointed in your readings of the current Ayurvedic literature. Given the lack of understanding in conventional medical (physiological) terms of where *doshas* originate, the allopath remains skeptical.

About homeopathy, the conventionalist is scornful. In some of the medicinal dilutions that homeopaths prescribe, not even one molecule of the plant or other substance from which the medicine was derived is present. According to conventional thinking, the homeopathic medicine must then be regarded as pure "placebo"—an intake of sugar pills disguised as medicine—and the cure must be considered entirely fortuitous.

In the same vein, spiritual healing, the idea of relying on Spirit for healing, encounters resistance. The Spirit, for an allopath, is a dubious concept, and, therefore, relying on it is tantamount to relying on the natural processes of the body, which are often inadequate for healing. To do so when all the powerful drugs of allopathy are available seems preposterous to allopathic practitioners.

Many practitioners of alternative medicine are equally scornful of allopathic practice. Allopathic drugs are mostly poisons to the body with harmful side effects, they say, so why should we poison the body when alternatives are available? For chronic and degenerative diseases, allopathy is ineffective anyway. Finally, allopathic medicine is not cost-effective. As the reader surely knows, it is the economics of allopathic medicine that are making people look for alternatives in medicine.

How do we go from these deep divisions among the practitioners of the two camps to an Integral Medicine that both camps can accept? The answer is this: We have to go to the philosophical roots of all medical practices and discover the unifying bridge-building philosophy.

The Disparate Philosophies behind Conventional and Alternative Medicine Practices

Our normal tendency is to see things as separate, and the scientist's job is to discover the thread that unifies, that weaves the separate flowers into one unified garland.

The various alternative medicine practices sound more mysterious than they really are. This is because their practitioners have tacitly been sold on the universal validity of the materialist metaphysics. According to this metaphysics, all is made of matter and its correlates, energy and force fields. All phenomena (this includes what we call mental and subtle energy, even what we call Spirit) are due to elementary particles and their interactions at a submicroscopic level.

In this model, causation is always upward causation, always rising up from the base level of elementary particle processes (see figure 1). Elementary particles make atoms, atoms make molecules, molecules make cells (some of which are neurons), and neurons make the brain. The cells make all the energies of the body that must include the subtle energies of alternative practices if these energies exist. The brain makes the processes that we call mental or spiritual.

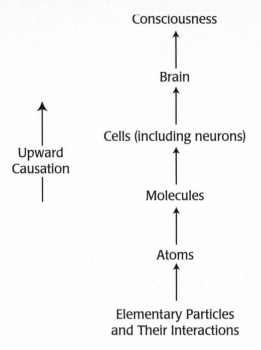

Fig. 1. The upward causation model. How upward causation works: Elementary particles make atoms, atoms make molecules, molecules make cells (neurons), neurons make the brain, and the brain makes consciousness. Causation rises from the base level of the elementary particles upward. Only elementary particles, nothing else, have causal efficacy.

In this view, to ascribe causal efficacy to an upper level of the hierarchy in which matter exists is paradoxical. How can the mind, an aspect of the brain, have causal efficacy of its own to influence the brain to produce a cure? It seems like brain acting on brain without a cause—a paradox.

In the same vein, to the materialist, the subtle energies of traditional Chinese medicine must be products of the underlying chemistry of the body cells. But then, how can a by-product of the cells or their conglomerates, the organs, produce a cure for what causes that by-product? Paradox again.

How can mere faith (because a "doctor" said so!) make sugar pills curative, as in homeopathy? Paradox, once more.

But if you are a reader of the history of the alternative healing literature, it won't take you long to observe that there are several paradigms operating there. The vast majority of the healing literature in the West is of course materialist—based on materialism, which is the conceptual foundation of conventional medicine.

In the alternative medicine paradigm, you can see three primary currents: One is based on the idea of "mind over body," that mind causes disease and mind heals, hence mind-body healing. Mind over body is possible because the causally efficacious mind is nonphysical. Mind is not brain. The Freudian idea of disease as suppressed emotional thought and healing as the awareness of the suppression (Sarno 1998) falls in the same category—psyche over soma, mind over body.

The second current is based on the idea that a nonphysical "life force," variously called subtle energy, *prana,* or *chi,* is the causal agent behind healing. Subtle energy is not a by-product of material chemistry; instead, it is the movement of a vital world. Hence a more appropriate English word for subtle energy is "vital" energy. Eastern models of healing, both Chinese and Indian, fall into this category.

A third current is the idea of a nonphysical Spirit (or God) who is the healer in all cases of spiritual healing. Spiritual healing is "God's Grace." Here one makes the distinction between "self healing" (healing when only you are involved and nobody else) and "other-healing" (healing with the help of somebody else, a healer). But in either case, the ultimate causal efficacy is given to a nonphysical entity called God (or Spirit).

To the materialist, to the practitioners of conventional medicine, a causally efficacious nonphysical mind, nonphysical subtle energy, or nonphysical God is dualism. And dualism is scientifically untenable. This has been argued since the time of Descartes, who tried to introduce mind-matter dualism. If mind (or subtle energy or God) and matter are dual substances, having nothing in common, how can the two interact without a mediator? You may have read the book *Men Are from Mars, Women Are from Venus,* by John Gray. It is about the difficulty men and women have in communicating because they don't have much in common. Fortunately for them, the sociologist or a counselor can act as the mediator. But where is the mediator to mediate the interaction of mind and matter, subtle energy and the body, God and the world?

Also, if you assume that mind and matter somehow interact anyway, experiments seem to rule that out. Such interactions would require exchange of energy between the physical and the mental domain. But it is an experimental fact that the total energy of the physical domain never changes!

Yet implicit dualism exists even in materialist thinking. This becomes obvious when one ponders such questions as, What causes the subject-object-split awareness that we experience? As the philosopher David Chalmers has cogently argued, this hard question cannot be answered within the materialist dogma. Materials are objects; interaction of objects always produces other objects, never a subject that experiences an object separate from itself. So the subject implicitly remains a dual (nonmaterial) entity even in materialist thinking.

One of the further irritating (but true) characteristics of all of the above systems of healing is that each one, conventional or alternative, is seen by its practitioner as exclusive—it alone can explain all healing. For example, many who believe in spiritual healing think like Mark Twain, who said, "God cures, and the doctor sends the bill."

But if you want to build an inclusive paradigm, how do you join contradictory philosophies? You need an inclusive philosophy. This inclusive philosophy—a metaphysics based on the primacy of consciousness in this age of science—is the gift of quantum physics and is the philosophy of the quantum doctor. In quantum physics, what we normally perceive as "things" are seen to be not things; instead they are seen to be *possibilities for consciousness to choose from.* This single idea has the potency to integrate all the disparate philosophies underlying the different schools of medicine. And more. It has the potency to validate your particular search for healing and show you how to achieve it.

2

My Story: How a Quantum Physicist Came to Meddle in Health and Healing

I was not always an integralist. In the mid-1970s, after I read Fritjof Capra's *The Tao of Physics,* a book on the parallels of science and spirituality, I began looking for a new paradigm of science that integrates science and spirituality. The mere existence of parallels did not satisfy. Also, it did not make sense that there should be one worldview for science and another for spirituality. By 1985, I had a strong intuition that an idea of the mathematician John von Neumann (1955) about quantum measurement was important. According to von Neumann, when we measure a quantum object, consciousness changes the quantum waves of possibility of the object into an actual event of experience. I thought this could be the basis for the new integrative paradigm of science.

Unfortunately, I was still convinced that everything is made of elementary particles, and von Neumann's idea was not compatible with my materialist prejudice. If consciousness itself is made of matter, is a mere epiphenomenon, a secondary phenomenon, of matter, then how can it causally act on matter?

There is, of course, an alternative to material realism (that is what the philosophy of objective independent separate material reality is formally called, which we have abbreviated in the preceding pages as materialism): mind-matter dualism. But as I mentioned before, dualism is fundamentally nonscientific.

So I was stuck like many others in the Never-Never-Land of materialism and dualism when lightning struck. I was hobnobbing with mystics at the time, secretly hoping to get some new insight for my research in spiritual practices. I was visiting a friend in Ventura, California, with the agenda of attending a Krishnamurti talk in nearby Ojai. After the talk, we were settled down in my friend's living room with a mystic named Joel Morwood. The conversation heated up in no time. I was a little high-handed with Joel in explaining the latest in New Age science—how paradoxical it was that consciousness, no doubt an emergent phenomenon of the brain, nevertheless collapsed quantum possibility waves of all the objects we see, including those in the brain.

And Joel challenged, "Is consciousness prior to the brain, or is the brain prior to consciousness?"

By then I was quite aware of the mystical ontology: Mystics put consciousness prior to everything. So I recognized the trap and said, "I am talking of consciousness as the subject of experiences."

"Consciousness is prior to experience; it is without an object and without a subject," said Joel.

Just a few months before, I had read a book by the mystic-philosopher Franklin Merrell-Wolff entitled *The Philosophy of Consciousness without an Object*. So I said, "Sure, that is vintage mysticism, but in my view you are talking about the nonlocal aspect of consciousness."

It was then that Joel gave me an emotional lecture about how I wore scientific blinders. He ended with the Sufi statement, "There is nothing but God."

Now mind you, I had heard or read those words many times by then, but this umpteenth time, understanding dawned, and a veil lifted. I suddenly realized that the mystics are right, consciousness *is* the ground of all being, including matter and brain, and that science must be built on this metaphysics rather than on the traditional materialist metaphysics.

I spent the next few years building this new science in a leisurely fashion. Actually, I am still working on it. In this book, I am sharing with you some of the exciting findings of the integrative potential of this new *science within consciousness,* in the field of health and healing.

How did I get into health and healing in particular? In the summer of 1993, I was invited to give a talk at a biology conference, an informal conference arranged by the University of California at Berkeley biologist Richard Strohman. At dinner one evening, a young biologist asked me what the selling point is of the new paradigm of science based on the primacy of consciousness that I was proposing. "Well," I said, "the paradigm integrates physics with psychology and spirituality. Conventional science is a science of objects; it develops theories of objects in terms of other more fundamental objects. So it fails for consciousness because consciousness consists of both subjects and objects. The new paradigm treats subjects and objects, spirit and matter, on the same footing."

"That's too esoteric," dismissed my young friend. "The average person today is not interested in the integration of physics and psychology, even spirituality. What else have you got?"

"The old science is about only the conditioned behavior of the world; the new paradigm also can handle the world's creative aspects. Hence it allows us to explore new avenues for our creativity. Surely, everyone is interested in creativity," said I.

"Maybe," my young friend allowed noncommittally. "What else have you got?"

Suddenly I understood what he was getting at. "You mean, is there an application of the new paradigm that is so striking that it will grab public imagination? There is. The new way of doing science not only may answer some knotty problems of biological evolution [the emphasis of my talk at the conference]. I am realizing that it may also be able to integrate the disparate ideas of conventional and alternative medicine," I said, a little out of breath as the sudden intuition came to me.

"That's the one; that's your communicating point to the public," said my new friend with enthusiastic approval.

Well, from there to here, to communicating with you the fruits of my attempt at integration, was a long journey. Fortunately, I did not have to start from scratch. I had developed an interest in health and healing long before. When I was growing up in India, I heard and read many stories about yogis with unusual control over their body functions. I also heard many stories about unusual healing powers. In fact, I witnessed a demonstration of this.

When I was in my early twenties, my teenaged younger brother had a severe stomach ulcer. If he ate anything even remotely spicy, he would get spasms. We had a boys' club of sorts, and a *sadhu* (a wandering, orange-robed renunciate, a common sight in India) started coming to our club to talk to us about spirituality. It was a nice break from politics and economics. But, of course, we were all materialists and gave the *sadhu* a hard time; my brother especially did so.

One day he challenged the *sadhu* to stop talking unless he had a demonstration of spiritual power for us. The *sadhu* thought in silence for a few moments and asked my brother quietly, "I have heard that you have an ulcer. Is this right?" My brother nodded. The room became very quiet as the *sadhu* gently laid his hand on my brother's stomach for a few seconds, then said, "Your ulcer is cured." My brother, of course, was not going to take his word. He immediately proceeded to eat a heavily spiced meal, but no pain erupted. His cure was genuine.

This incident created in me an interest in the subject of health and healing, one that persisted over the years. Another lasting impression was created by homeopathy. As you probably know, homeopathy was discovered in the West but is practiced more in India than anywhere else. When I was a child, my family used to eat a lot of fish. Invariably, every now and then, I would get a fish bone stuck in my throat. On such occasions, my mother always gave me a dose of a homeopathic medicine, Sulfur-30. Within a couple of hours, the fish bone would go down and comfort would return to me. My child's mind was quite impressed with this and other homeopathic remedies.

Growing up in India, I was naturally quite familiar with Ayurveda. I don't remember at this late date any particular Ayurvedic remedies that I may have used in my childhood, but I do remember one thing. In Bengal (the state of India where I lived as a child), an Ayurvedic physician was called *kaviraj,* meaning "king of the poets." It intrigued me that healing can be regarded as writing poetry!

So going back to the night when I had that conversation with the young biologist, as I was returning to my hotel, the memory of an Ayurvedic physician being called the king of poets came back from across many decades. I found myself thinking, is healing poetry? Certainly, the way it is practiced in the allopathic tradition in which materialism and reductionism reign, it is very prosaic. You go to a doctor's office, machines measure your various health indicators, the doctor looks at the machine readings, and only then is he ready to help you. No poetry there. The help you get is also mechanical. Just like classical physics, all is routine, all is determinism.

By contrast, alternative medicine traditions are fundamentally a little subtle, a little vague. The vital and mental bodies the doctors talk about are subtle, but they don't mind. What they do is often not quantifiable, but they are at ease with it. They often use intuition for their diagnosis, not machines. But they are okay about it. Their language of communication is vague, but they manage. What they do is a lot like poetry. I realized it is that old battle between art and (deterministic) science after all.

And you know what? I will tell you the fundamental secret of quantum physics, the reason materialists have such a hard time understanding it. Quantum physics is also a lot like poetry. Instead of determinism, quantum physics talks of uncertainty. Instead of particles *or* waves, this *or* that outlook of classical physics, quantum physics introduces complementarity, wave *and* particle, this *and* that. Most important, quantum physics brings consciousness into physics: *who* is looking into the affairs of what is being looked at. Can you imagine talking about poetry without talking about the poet?

Yet quantum physics explains a lot of prosaic experimental data. In this way, quantum physics has the potential to combine both art and deterministic science, creativity and fixity. I think it is the appropriate vehicle to integrate the "poetry" of alternative medicine and the "prose" of allopathy.

I was elated, ready to apply the new science within consciousness to integrate conventional and alternative medicine. But it took ten years from there to here.

The Conceptual Blocks

Actually the conceptual basis for the paradigm shift that I develop in this book has been around for some time. But it involves a radical shift of our worldview, especially the worldview in the West, and this has been the stumbling block.

For example, what is lacking in the field of mind-body medicine is a clear understanding of the manner in which mind and body can interact without dualism and of where the causal efficacy of that interaction lies. In other words, we have to be looking for an understanding of what psychologist Donald Campbell calls *downward causation,* popularly thought of as "mind over matter," but without dualism. The absence of such understanding is clearly impeding the future progress in this field. Most researchers are not aware that their classical physics prejudices prevent them from seeing the solution here. I notice enormous inertia, huge reluctance to change from the classical to the quantum worldview, and this resistance comes from practitioners of both camps of medicine.

In a classical worldview, downward causation is either a paradox or it leads to dualism (Stapp 1995). In the quantum worldview, downward causation is a *fait accompli.* In order to make the quantum worldview consistent and philosophically astute we have to introduce downward causation with consciousness as its agent. Quantum physics looks at things and their movements as possibilities. *Who* chooses from amongst these possibilities? After more than seven decades of research, the only logically consistent answer that has emerged is this: consciousness chooses. And therein lies its power of downward causation.

Let's consider the other mainstay in alternative medicine: Eastern medicine as practiced in China, India, and other Asian countries. Easterners, in their healing practices, use such concepts as the flow of *chi* in Chinese medicine or *prana* in Ayurveda. When you read ancient treatises it is clear what *chi* or *prana* is—some sort of energy, but nonphysical. But when you read more modern expositions of these subjects, especially in the West, my experience is that nobody is very clear on what *chi* or *prana* means. It is left vague whether *chi* or *prana* is a physical or a nonphysical entity.

It should be clear to you that most modern authors are pussy-footing around the meaning of these concepts because they cannot explain them in terms that are acceptable to the modern scientific worldview.

There is actually a corresponding concept called vital energy in the West, but it conjures up the image of dualistic vitalism, a philosophy that biologists discarded some time ago. So in general, researchers and scientists in the West, even the aficionados of alternative medicine, are reluctant to use the phrase "vital energy." Instead they opt for "subtle energy" and most of them continue their materialist beliefs about what subtle energy is.

Some look at subtle energy as an emergent phenomenon of the living cells and organs of the body. Others think of subtle energy as energy of a higher frequency than gross energy. Other more maverick researchers explore the idea of an "electromagnetic body" overlaying the usual "biochemical" body to explain the intricacies of subtle energy.

But one has to go no further than to consider homeopathy to convince oneself that there have to be nonphysical agents at play in healing, at least in some cases of healing. In homeopathy, a medicinal substance is applied (orally) in such dilute proportion that scientific calculations show without a doubt that at most potencies not one molecule of the original substance may find its way to the seat of the disease. And yet the success of homeopathy seems to hold up, even in double-blind clinical trials. And oh yes: These trials prove that homeopathy is not a placebo (sugar pills). So if the efficacy of homeopathy is true, then there must be agents of healing that are nonphysical! We have to come to terms with this idea of nonphysical agents of healing.

Samuel Hahnemann, the German discoverer of homeopathy,

suggested in the early 1800s that there are nonphysical vital energies that operate in homeopathic healing. He used the term *dynamis* to denote vital energy.

In this way, I realized very soon after I started my research that alternative healing practices remain a mystery (and therefore controversial) to most people of conventional thinking because their proponents suffer from five philosophical shortcomings:

1. They fail to distinguish between mind and consciousness. Long ago, Descartes put the two concepts, mind and consciousness, together as the one concept of mind, and that error still haunts medicine.

2. The causal role of consciousness as the origin of downward causation is either missed or is obscured in ambiguity. Somehow the lessons of quantum physics have not penetrated the classical physics armor of even the practitioners of alternative medicine.

3. The distinctive role of mind as opposed to the brain is missed. Scientific progress in this field already a decade old has been missed.

4. The distinctive role of the vital body compared to the physical body is also missed. Recent scientific progress has been missed here also.

5. Neither consciousness, nor the mind, nor the vital body is acknowledged to be nonphysical. We have to solve the problem of dualism, but who says that there is no way to get around dualism, that it is an insurmountable problem?

It is through solving these knotty problems of philosophy that I arrived at a science within consciousness for medicine, or what I call Integral Medicine. This paradigm acknowledges and includes the rediscovery of downward causation by consciousness in quantum physics. It also takes advantage of the rediscovery of the mind and the vital body within science. It then uses quantum thinking as a way of introducing the mental and vital bodies as distinct from the physical and without falling prey to dualism.

Applying the paradigm to actually explain how the various systems of alternative medicine work was fun for me and often very inspiring. It was also humbling at times because I knew so little about so many things in medicine. Even now I feel that I have hardly scratched the surface of the enormous explanatory potential the new paradigm brings forth.

The Plan of the Book

Part 1 is the survey of Integral Medicine and the rest of the book is the application of the model. Chapters 3 through 7 resolve the philosophical issues, give you an adequate dose of quantum physics (enough to be fun, and definitely not too much to make you uncomfortable), and set up the basic paradigmatic framework. Once you read part 1, you will be equipped with proper philosophy and the subtleties of quantum thinking and ready to consider applications to all the different systems of medicine.

Part 2 is the subject of vital body medicine—Ayurveda, traditional Chinese medicine (including acupuncture), chakra medicine, and homeopathy. Where do the Ayurvedic *doshas* originate? Why do we catch a cold when the season changes? What is vital energy? How can we learn to explore vital energy? Why don't we find either it or its passageways in the physical body? Why does the acupuncturist put needles on the arm to affect the lung? Are the Eastern concepts of the chakras for real? If homeopathy is not a placebo, how does it work? These are some of the questions we address in part 2.

There is more here on the vital body and its quantum nature, which you will need to appreciate the poetry of Ayurveda and traditional Chinese medicine, the chakras, and the subtleties of homeopathy, although the details look a lot more like prose. This part ends with homeopathy—the ultimate of subtlety in which less is more.

In part 3, I deal with the subtleties of the quantum mind, explain how the mind creates disease, and explore some of the existing techniques of mind-body healing. Here the emphasis is on understanding emotions, how mind imposes itself on feelings, how that produces disease, and how we can control the mind, prevent mind-body disease, cope with it, and even heal ourselves using the techniques of mind-body medicine.

Part 4 discusses mainly the spiritual underpinnings of healing, seeing healing as an opportunity for spiritual growth. I hope you will be properly inspired to look at a new kind of intelligence as part of your healing journey. Can we heal ourselves? Yes, provided we approach healing creatively as art. The mysteries of spontaneous healing are also fully explained. The mind's new science as developed here is shown to resolve the paradoxes of mind-body healing such as: Who is healing whom? Why do some people heal and others don't using the same techniques? What can we do to maintain health at all levels of our being? Can we move from the current preoccupation with disease to a preoccupation with positive health?

I end the book with a speculative chapter on the ageless body.

The book as a whole teaches you how to explore health and healing in a quantum universe, our universe. As such, the book is a series of enablers along these lines:

• Quantum and primacy-of-consciousness thinking enables you to look at your health in an integral, truly holistic way, with an integral philosophy as your guide (see chapters 3, 4, and 6).

• The book enables you to think of disease and healing in terms of a useful new classification. This classification enables you to know when to apply what form of medicine—conventional or alternative; and if alternative, which specific form of it (see chapter 4).

• This new approach to medicine enables you to see clearly that you have the power to choose between disease and health (see chapter 6).

• Quantum thinking even clarifies the role of allopathic medicine in health and healing (see chapter 7).

- Quantum and primacy-of-consciousness thinking clarifies and explains for you many unexplained mysteries of Eastern medicine, homeopathy, and mind-body medicine (see parts 2 and 3).

- In particular, the theory developed here enables you to understand the nature of pain and to adopt strategies to deal with it (see chapters 14 and 15).

- The book enables you to understand the meaning and context of disease and illness and may inspire you to contemplate undertaking the journey of self-healing if ever you need it. And if you do undertake such a journey, the theory here gives you guidelines (see chapters 16 and 17).

- Quantum physics thinking gives clear guidelines for doctor-patient relationships (see chapters 6 and 16).

- Most important, the goal of the book is to enable you to see illness and healing as integral parts of your exploration of yourself, your particular search for wholeness, and finally to empower you to choose wellness over illness, and wholeness over separateness.

Have I succeeded? You be the judge.

3

The Integration of the Philosophies

Most medicine practitioners are believers in classical physics, the seventeenth century deterministic physics that Isaac Newton built. According to this physics, all movements are material movements and are determined by physical laws and the initial values of position and velocity of the material objects concerned.

But in the 1920s, physics underwent a monumental shift from classical deterministic physics to a new physics. In this new physics, called quantum physics, objects are described as waves of possibility that can be at two (or more) places at once. But in which of these places an object will manifest in a given measurement is not determinable by any physical law or algorithm.

If this indeterminacy is not enough, there are other striking differences between classical physics and quantum physics. In classical physics, all interactions are local, coming from the vicinity with the help of signals traveling through space taking a certain amount of time. But in quantum physics, this is not so. There, nonlocal connections that allow signal-less instantaneous communication exist in addition to local connections. In classical physics, all movement is continuous and hence determinable by mathematics and algorithms that require continuity. But in the new physics, discontinuous "quantum leaps" are allowed in addition to continuous movement.

Thus quantum physics is often looked upon by classical physics aficionados as magical and mysterious, not to be trusted except as an instrument for explaining data and predicting experimentally verifiable results. Those who look upon quantum physics this way are strict materialists (matter is the only reality) at heart and reductionists (everything can be reduced to elementary particles of matter and their interactions) in their methodology. But if you are an integralist looking for a new organizing principle (consciousness), then you want to look at the "magic" of quantum physics and wonder how it can be like that. You question the validity of your materialist worldview.

In quantum "magic," you will discover a rationale for consciousness and downward causation outside the jurisdiction of the materialists' upward causation. Consciousness causally acting upon matter, consciousness over matter (downward causation), now becomes a potent idea because you intuit the importance of this idea in biology, psychology, and especially medicine.

Classical physics forces us to look at objects as "things" whose movements are completely determined by the laws of physics and some initial conditions (initial position and velocity). In contrast, in quantum physics, objects are calculated as waves of possibility, not as determined movements. It is observation by an observer that *precipitates* a definite event out of the various possibilities. Thus the visionary window opens: The possibilities are possibilities of consciousness to choose from; when consciousness chooses, an actual event consisting of a subject looking at an object precipitates. This is what the quantum physicist calls the event of "collapse."

Don't be confused by the word "collapse." Collapse means simply the change from possibility into actuality. Upward causation due to elementary particle interactions gives us possibilities; it takes a nonmaterial consciousness to cause collapse, to choose actuality from possibility. To avoid confusion, wherever necessary, I will use the phrase "quantum collapse" instead of just "collapse."

This collapse is the causal power of consciousness that we are looking for; it is downward causation (Goswami 1989, 1993).

This also sounds like dualism until you make a radical shift about the conceptualization of consciousness. Consciousness cannot be made of elementary particles; it cannot be a brain product. But it cannot be a dual separate object either. There is a third way of thinking about consciousness that you must comprehend. Consciousness is the ground of being. Understand that the material possibilities are possibilities of consciousness itself to choose from; they are not outside consciousness, not separate from consciousness.

Still confused? Look at the gestalt picture of figure 2, a drawing that has two pictures within the same lines, an old woman and a young woman. The artist called the drawing "My Wife and My Mother-in-Law." Initially, you see one of the two, either the old woman or the young. But keep shifting the perspective of your looking; soon you see the other. What is happening? You are not doing anything to the picture! You are *choosing* by shifting the perspective of your looking, of your recognizing. Collapse or downward causation is like this.

Fig. 2. "My Wife and My Mother-in-Law" (based on the original drawing by W. E. Hill).

So what materialists think of as an epiphenomenon is the real thing with causal efficacy. Thus looking at health and healing through the worldview of quantum physics immediately endows the healer and the patient with the power of downward causation, with the potential power of choosing health over disease. What remains is to learn the subtleties of the exercise of this choice. And when you do (see chapter 6), you find out why such self-healing is traditionally called a healing by a higher power, Spirit, or God: because it takes you out of your ordinary separate cocoon of existence to a holistic, nonlocal level of being.

Okay, so material healing is healing through upward causation, and spiritual healing is healing through downward causation. How do we incorporate mind-body healing or vital energy healing in our integrative paradigm?

When we experience material objects, we do it via sensing. Is sensing the only form of experience of which we are capable?

Psychologist Carl Jung codified the answer to the question above most succinctly (see figure 3). When Jung was researching personality types, he discovered that each of us dominantly uses one of four ways of experiencing: sensing, thinking, feeling, and intuition.

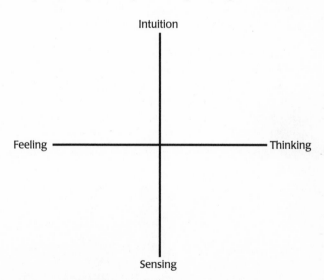

Fig. 3. Carl Jung's typology. The different ways we experience reality.

When consciousness collapses a material wave of possibility, we manifest the experience of sensing as part of our material body experience. So how does our experience of thinking arise? Thinking must be the result of collapsing a possibility wave of the mind—our mental body. Similarly, collapsing a possibility wave of subtle or vital energy, a movement of the vital body, gives us the experience of feeling. And intuition is the way we experience still another category of possibilities of consciousness—let's call it the supramental domain.

But materialists can object. We know that thinking most often involves memories that are stored—where else?—in the brain. So how do we know that thinking is not a brain phenomenon?

In the same vein, how do we know that the possibilities whose collapse gives us the experience of feeling do not belong to the physical body itself? Isn't feeling associated with the nervous system and the brain? Do we really need to postulate the vital body?

The Rediscovery of the Mind

When computer science was being developed in the late 1950s and early 1960s, one of the earliest ideas that caught the attention of the scientific world was that of artificial intelligence—building a computer that can think. Through the 1970s and 1980s, computer scientists continued to write programs of thinking that so nearly mimicked human thinking that they could fool many people. In the 1980s, you could call up a number in Canada and talk to a computer with a program that mimicked a California New Age psychotherapist. You were not able to tell that you were not talking to a counselor of that type, so equipped was the program with all the jargon of New Age psychotherapy.

So can computers think? If computers can carry on a conversation with us, thinking humans, how can we deny the capacity for thinking to computing machines? And then, since the brain is a computer, why should we doubt that thinking originates in the brain itself? A nonphysical mind is not needed.

Not so fast, cautioned the University of California at Berkeley philosopher John Searle, who in the 1980s argued against a thinking computer (Searle 1987). Computers are symbol-processing machines, said Searle. They cannot process meaning. Thinking involves processing of meaning. So computers do not think! Thinking requires a separate mental body. We have it, and this is why we can think. Searle (1994) later wrote a groundbreaking book appropriately entitled *The Rediscovery of the Mind.*

Computer scientists did not take Searle's ideas very seriously, but then the mathematician-physicist Roger Penrose (1989) wrote a book evocatively called *The Emperor's New Mind.* Penrose proved with the help of high-power mathematics that computers indeed cannot process meaning, just as Searle surmised. As the emperor's new clothes were false, so is the computer's purported "mind."

So mental possibilities are distinctly nonmaterial. They are possibilities of meaning. When consciousness collapses these meaning possibilities in conjunction with brain possibilities, the collapsed brain actuality makes a representation of the collapsed mental meaning of the experienced thought.

If material movements are possibilities, then it makes sense to posit that mental movements are also possibilities—meaning possibilities. When we choose from the meaning possibilities, we have a concrete thought. Consciousness, in every experience, not only has a physical perception of a physical object, but also a mental perception of its meaning.

Consciousness is not mind; it is the ground of all being, the ground of both matter and mind. Matter and mind are both possibilities of consciousness. When consciousness converts these possibilities in a collapse event of actual experience, some of the possibilities are collapsed as physical and some as mental.

In this way, consciousness clearly is seen as the mediator of the interaction between mind and body, and there is no dualism (Goswami 2000). And now room is made for mind-body healing in which a proper role is given to consciousness (the causal agent of downward causation) and to mind (from which meaning comes) in relation to the physical body and its healing.

We are also enunciating a new kind of psychophysical parallelism. The philosopher Gottfried Leibniz (1646–1716) proposed an alternative that he thought might avoid the dualistic pitfall of

Cartesian mind-matter interactionism. Mind and matter never interact, said he; they just function in parallel. But other philosophers did not particularly warm up to Leibniz's idea because of this puzzle: What maintains the parallelism? So Leibniz's philosophy, too, smacks of dualism.

Now, finally, after a few centuries, with quantum thinking, we see the resolution of the predicament of both Descartes' and Leibniz's philosophies. What mediates the interaction of mind and matter? Consciousness does. What maintains the parallel functioning of mind and brain? Consciousness does.

So from a quantum point of view, it is not hard to see why mind and meaning are important in medicine. Ordinarily, we live in a separate reality, separate from the whole of consciousness. It is our conditioning that gives us individuality. There is no one-to-one correspondence between objects and their meaning. It is only our conditioning that makes us think there is. And no wonder we are fooled into believing in independent separate objects; from this separatist point of view, we partake in actions that can either increase our sense of separateness (as when we assign limited meaning to our experience) or decrease it (as when we creatively expand the meaning). The former is suffering, of course, but we may not immediately see it. Disease is a reminder—like getting hit with a two-by-four—to change our ways and correct the direction of our journey toward wholeness, which is where healing takes us.

What Does the Vital Body Do That the Physical Cannot Do?

The rediscovery of the vital body occurred at about the same time that modern science was rediscovering the mental body in the 1980s. Here a crucial step occurred through the work of the biologist Rupert Sheldrake (1981).

Biologists had eradicated the philosophy of "vitalism," which postulates a nonphysical vital body as the origin of "life force," in the 1950s with the discovery of molecular biology, which showed enormous promise for an understanding of everything there was to know about life. Alas! The enthusiasm did not last long, as molecular biology could not explain the phenomenon of morphogenesis—how a single-celled embryo grows up to be a differentiated biological body of organs.

An embryo expands by cell division, making an exact replica of itself with all the same DNA, the same genes. But in the adult body, the cells are differentiated as to their functions. For example, the liver cell functions differently from the brain cell. The proteins a cell makes determine cellular function; the genes have the code to make the proteins. In liver cells, the genes are being activated so as to make an entirely different set of proteins from brain cells, so there must be programs running the cells. But the source of the programs is not part of the DNA.

So already in the late 1950s, biologists such as Conrad Waddington (1957) were postulating the idea of epigenetic morphogenetic fields that, perhaps residing in the cytoplasm, outside the cell nucleus, guided the programs of morphogenesis. But nobody ever found epigenetic guides of morphogenesis either.

Sheldrake (1981) proposed an explanation of the hitherto unexplained phenomenon of morphogenesis in terms of nonphysical and nonlocal morphogenetic fields residing outside space and time. This clarifies the role of the vital body that now can be seen to be the abode of the morphogenetic fields as distinct from the physical: The vital body provides the blueprints for the forms and programs of morphogenesis. The blueprints themselves are designed for vital functions, maintenance, reproduction, and so forth.

The philosopher Rudolf Steiner, as early as 1910, saw morphogenesis as the function of the vital body (which he referred to as the "etheric" body) and used this conceptualization as part of the formulation of his anthroposophic medicine (for a discussion, see Leviton 2000). Sheldrake's work confirmed Steiner's foresight. Yet, one asks, isn't it true that Sheldrake's theory of vital blueprints is far from being accepted in the biology mainstream? Isn't it also true that most biologists feel that a materialist explanation of morphogenesis is right around the corner?

So what else is new? Unjustified materialist claims continue to be a nuisance for the builders of the new paradigm. Pay attention to the conceptual consistency and clarity of the new paradigm ideas; for example, appreciate the fact that there are no paradoxes. Instead, the paradoxes of old paradigm thinking are being resolved.

As for Sheldrake's work, the question is not if other biologists are accepting it, but is it useful? I have found it enormously useful and already have employed it in explaining creativity in biological evolution (Goswami 1997, 2000). Here I use it in the integration of vital body medicine and physical and mental body medicine, no less. The concept of morphogenetic fields also helps us to understand the famous empirical concept of the chakras (see discussion to follow and also chapter 11). Ultimately, of course, it is a question of experimental data. It will take a while experimentally to check out everything about Sheldrake's theory.

Meanwhile, look at the enormous scope of the new paradigm, science within consciousness, of which Sheldrake's idea is part and parcel. It is explaining practically all the data that is anomalous from the point of view of old paradigm thinking (Goswami 1993, 1999, 2000, 2001; Blood 2001).

When consciousness simultaneously collapses the possibility waves of the physical body and the vital body, the physical makes a representation of the vital blueprint, to carry out the vital function of the relevant morphogenetic field in the physical world.

So organs are representations of vital body blueprints of various morphogenetic fields. It is well known that there are places in the physical body where we tend to feel vital energy most easily. These are called the chakra points. Many authors (Joy 1979; Motoyama 1981) have noted that the chakra points are found close to important organs of the physical body. Now we see why. These are the points where representations are made of the vital onto the physical. Once the representations (the organs) are made, the quantum collapse of the functions of an organ is always associated with the quantum collapse of the correlated vital blueprint. The activation of the vital blueprint is tantamount to vital energy movements. These movements are what we experience as feelings.

So the rediscovery of the vital body is giving us an understanding of another important phenomenon—the chakras. Chakras play an important role in Integral Medicine (see chapter 11).

The new vital-physical parallelism shows promise of integrating Western and Eastern medicine. Yes, the physical body chemistry is important, as is the hardware of the computer. So conventional medicine is not wrong. But also important are the correlated vital body movements that consciousness collapses along with the physical body representations and their functions.

In general, wellness requires the homeostasis not only of the physical body functioning—the functioning of the mapped blueprints of the vital body—but also of the vital body movements. Eastern medicine concentrates more on the lack of balance and/or harmony of the vital body movements.

What do imbalances of the vital body movements mean? In traditional Chinese medicine this is an imbalance of the yin and yang aspects of the vital energy *chi*, the particle and wave aspects of *chi*, if you will. This is vague, to be sure, but it gives us a big hint.

We operate in two different modes of self-identity—ego or the classical mode, and the quantum self or the quantum mode (Goswami 1993). In the classical mode, we are localized and determined; we can call it the particle mode of identity. In the quantum mode, we are nonlocal and free; we can recognize it as the wave mode. So balancing the modes of movement of the vital body means balancing the classical and quantum modes—the conditioned and creative modes, if you will—of self-identity in the operations of vital body movements.

In other words, a balance of conditioned (yin) and creative (yang) movements of *chi* is needed for the proper maintenance of health. A dynamic homeostasis is needed where creative forays outside the conditioned patterns of the vital body are allowed.

The Eastern traditions of medicine see illnesses (especially chronic illnesses) as due to the imbalance of vital body movements. In their schema, people can be born with certain imbalances in the manner that vital movements are employed. The basis of their thinking is the theory of reincarnation—rebirth of a part of one's vital and mental essence (popularly called karma) in another physical body in another place and time.

The proof of the pudding is in the eating. Is reincarnation provable by science? For that matter, is even the talk about survival of our mental and vital bodies after death scientific? These are scientific issues because the scientific validity of survival-after-death and reincarnation also validates the existence of our vital and mental bodies.

Reincarnation used to be considered an Eastern concept not to be taken seriously by the more scientifically minded West. Thanks to the extensive new data on several fronts, this is no longer the case. First, the University of Virginia psychiatrist Ian Stevenson (1987) has published extensive data on both Eastern and Western children's reincarnational memories that he has empirically verified. Second, a new form of psychotherapy, in which the therapist provides therapeutic release by making unconscious reincarnational memories of the clients conscious, has met with much success. Third, reincarnation provides the most straightforward explanation of the phenomena of child prodigies, geniuses, and people's search for meaning in their lives.

There is also considerable data on near-death experiences (see, for example, Sabom 1982) that independently corroborate the phenomenon of survival after death. Clinically brain-dead people (whose EEG recordings are flat), when revived, report out-of-body experiences, the experience of going through a tunnel and meeting long-lost relatives, the experience of seeing a lighted being, life-review experiences, and so forth.

Are the phenomena of survival and reincarnation provable by science? Empirical data indicate they are. Moreover, great progress has been made toward building a proper theory of how the surviving vital and mental aspects of an individual are able to carry the individual signature (karma) from one incarnation to another (Goswami 2001).

If vital body movements are employed in an unbalanced way to begin with (due to vital body karma), with more yin or more yang, they will produce faulty functioning in the physical living body. If further vital body imbalance is produced in this life because the root cause of illness is not being addressed, then there will be further lack of synchrony between the correlated vital and physical states and the functions they perform.

When you have a physical ailment such as a headache, the Western physician looks for analgesics that will relieve the symptom, the pain. The Eastern acupuncturist, on the other hand, will try to find a way of correcting the particular imbalance between yin and yang functioning of the vital body that is the root cause of the pain. So the acupuncturist empirically discovers the particular point of the physical body to probe with the acupuncture needle to disturb the faulty conditioned movements of the vital body. The acupuncture probe is designed to be a trigger of the correctional mechanism for the vital imbalance.

If you are suffering from fatigue or a lack of vitality, Western medical practitioners will look for a cause such as anemia or hypoglycemia and treat the symptoms after they are clearly diagnosed. But if you go to a doctor trained in the Ayurvedic tradition, this doctor will treat you with herbal medicines that are designed to correct your *pranic* imbalance. Through empirical research and experience, the Ayurvedic physician knows the herb or combination of herbs most likely to help restore the balance of the *pranic* movements needed for the cure of your particular ailment.

In summary, Eastern medicine has concentrated on half the apple, the vital body where the blueprints of form are; Western medicine has focused on the other half, the physical body, the form itself. In this way we have two systems of medicine, both very good at what they do, but alone neither is the perfect apple of holistic health that holds the key to cures. So we must integrate them. This is the job of science within consciousness, and in this light, Integral Medicine is the truly holistic medicine.

4

Levels of Disease and Levels of Healing

Thinkers, listen, tell me what you know of that is not
    inside the soul
Take a pitcher full of water and set it down on the water—
    now it has water inside and water outside.
We mustn't give it name,
    lest silly people start talking again about the body
    and the soul.

*   —Kabir, mystic poet (Bly 1977)*

What is Kabir trying to say in this poem? That it is all con-sciousness, both body and soul. The difference between the water outside and the water inside arises from the glass boundary of the pitcher. The difference between body and soul arises from the dif-ferent ways we experience them: We experience the physical world of body as external to us, but we experience also an internal world of awareness, which, then, we call the soul.

On closer examination (as seen in the previous chapter), the soul (or subtle body or psyche) is found to consist of three bodies: the vital energy body, the mind, and the supramental (which we will also call the supramental intellect). Adding the physical and the unlimited whole (the ground of being, also called the bliss body), we have altogether five bodies corresponding to five worlds of consciousness.

It is interesting to go back to the Upanishadic story (Nikhilananda 1964) that tells us about the five compartments of reality, each beyond the other in terms of subtlety. The son of a sage, on being prodded by his father, is meditating on the nature of reality. He meditates and finds that without food, reality cannot exist, and we die. So he tells his father his realization: food (physical) is reality. (This is exactly what today's materialist scientist discovers.) The father says, "That is correct, but investigate some more." (Alas! There is no one to guide today's materialists.)

The son goes back to meditation again and goes deeper. He discovers the vital blueprints beyond his physical body. He experiences the feelings of the movement of these blueprints, the vital energies that are the feelings of aliveness. So he says to his father, "Reality is vitality, vital energy." The father says as before, "Yes. But go deeper." (Today's materialist of course denies the reality of vital energy, because he or she thinks that its conceptualization has to be "dualistic.")

The son meditates for years and realizes that without mind to give meaning to the vital energy feelings and the physical universe, reality is meaningless. So he says to his father, "Reality is mind." The wise father says, "Yes. But investigate even further." (Today's materialist gets stymied thinking that mind is brain because what else can it be? And because if mind is separate, what mediates the interaction of the mind and the brain?)

The son now meditates hard and goes deep inside his psyche. There he discovers the supramental body of contexts, the body of the laws of mental, vital, and physical movements that govern all changes of all the stuff of which reality is made. So he declares, "Reality is supramental intellect which governs all the other worlds." The father says as before, "Yes. That is correct. But go even deeper." (Materialists often wonder why mathematical laws govern the movement of physical objects; where do they come from? Some try to derive the mathematical laws of physics from the random motion of material substratum. If they paid attention to Plato's philosophy of archetypes instead, they could discover the supramental.)

Finally, the son meditates and discovers the wholeness of consciousness, the unlimited aspect that we experience as bliss. So he says to himself, "Reality is bliss." This time he does not go back to his father. He knows and understands everything. (Today's materialist scientists shake their heads and declare that meditation must be causing this fellow his hallucinations of bliss. Since everything is the movement of matter and matter exists in both order and disorder, how can the ultimate reality be only order, only bliss?)

Evidence and arguments are accumulating in favor of the five worlds of consciousness (some of it was already discussed in chapter 3). Rupert Sheldrake (1981) has shown how nonlocal and nonphysical morphogenetic fields are essential to understand biological form-making from the one-celled embryo. The instructions of form-making, cell differentiation (all cells contain the same genes, yet toe cell genes are activated very differently from brain cell genes), are nowhere to be found in the physical body, and that includes the genes (which are more or less instructions for protein-making). Could it be that the vital body is the reservoir of the morphogenetic fields?

Chinese acupuncture, Chinese herbal medicine, the Indian medicine of Ayurveda, and homeopathy all use the concept of vital energy (or *chi* or *prana*) as the agency that is implicated in the healing of the physical body. Could it be that they are talking about the modes of movement of the vital body morphogenetic fields? The answer is affirmative (see chapter 3, where I show that these vital energies are what we feel directly when we use the term "feeling").

What validated the mind? A tremendous step is the change of mind (pardon the pun!) of some computer scientists about the computer's ability to process meaning (see chapter 3). First, in the late 1980s, it was the philosopher John Searle, a critic of artificial intelligence, who started making noise about the computer's inability to process meaning. His argument was basically that computers are symbol-processing machines. If we reserve some of the symbols for processing meaning, we need to reserve more symbols to process the meaning of the meaning symbols, and then ever more symbols for processing the meaning of the meaning of the meaning symbols, ad infinitum. So there are never enough symbols for the computer to succeed in processing meaning.

How do *we* process meaning? We use our mind. The mathematician Roger Penrose gave a rigorous proof of Searle's idea using Goedel's incompleteness theorem. And now even artificial intelligence researchers (Banerji 1994) are following in Penrose's footsteps. So finally, the concept of the mind being needed as a separate nonmaterial world to process meaning is making sense.

What is the evidence for a separate body of supramental intellect aside from our puzzlement about the origin of the relation of mathematics and the laws of physics? When we investigate creativity (Goswami 1999), we find that creativity at the low level consists of finding new meaning, a shift of mental meaning from a conditioned old one to an inventive new one. This is called *situational creativity*.

But at the highest level, creativity consists of discontinuous leaps in the context of thinking itself. This is *fundamental creativity* and consists of discovery because here we are discovering the fundamental laws of movement of the different worlds—already present in the compartment of consciousness called supramental intellect, which we have forgotten and can access only through intuition. In contrast, situational creativity is invention, accessible, at least in principle, to reason. Invention depends on the discoveries of fundamental creativity, but not vice versa. So the existence of fundamental creativity points us to the existence of the supramental intellect world. Note, however, that the supramental world is the reservoir not only of the contexts of mental meaning but also of vital functions and physical laws.

Is there any new evidence of bliss outside of meditation? There is. People have discovered blissful states of *samadhi* under psychedelic drugs, holotropic breathing (Grof 1992), even in near-death experiences (Moody 1976). These experiences are validating the ancient (Indian) description of consciousness as existence-awareness-bliss. Of this trio, existence is the most immediately obvious. Most of us do not deny awareness (except maybe the philosopher Daniel Dennett, who thinks we are zombies!), but high levels of bliss used to be somewhat removed from everyday experience. Not anymore.

To summarize, consciousness has five compartments or bodies:

• The physical, which is the hardware and where representations are made of the subtler bodies.

• The vital, which carries the blueprints of biological functions that are then represented in the physical as the different organs.

• The mental, which gives meaning to the vital and the physical and of which the brain makes representations.

• The supramental intellect, which provides contexts for mental meaning and vital functions and associated feelings, and the laws of physical movement.

• The bliss body, which is the unlimited ground of being. In this ground of being with unlimited possibilities, the other four compartments exert progressive limitations.

Answer to Dualism

But what about dualism? This I already discussed in the previous chapter, but because of the importance of this issue, here is another exposition of it.

Dualism is primarily a problem of communication between substances that have nothing in common. Mind-stuff is the antithesis of matter-stuff. Mind acts nonlocally, has no extension in space-time, and cannot be quantified. Matter acts locally, has extension in space-time, and can be quantified. So the core of the problem is obvious as it is realized that these two substances can never have a way to communicate with each other. Yet our experience continuously shows that they can and do—you see an object and the mental meaning of it simultaneously arises in consciousness. How is this possible?

In order for anything to be, it must first be *possible* for it to be. If the possibility does not exist, then it is impossible for the actuality to exist. It is a necessary condition. This is good, basic logic. In looking at what this world of possibilities is like, it is more like mind-stuff: It has no extension in space or time; it cannot be quantified. It is also part of the ground of all being. This realm of possibility, then, can be equated to the earlier described bliss body of consciousness.

When the quantum collapse occurs, the result in the actuality of our experience is that of distinctive compartments (physical, vital, mental, supramental), which are seemingly unrelated to each other in terms of their "substance." You perceive the physical as external to your awareness and all the other compartments as internal. Among the internal compartments, you experience the mental compartment of meaning most easily. But we can learn to feel vital energy movements, not only our own but also those of others. And we can discover supramental intellect when we take a creative leap of discovery or in merely conceptual thinking by extrapolation. (What are concepts but conditioned contexts of meaning?)

Dualism no longer poses a problem because all bodies are possibilities in consciousness (in its bliss body) prior to collapse. When the quantum collapse occurs, it does so on all four levels and is thus applicable to each distinctive compartment or body of possibilities. Just as it is incorrect to say that blue creates yellow or red arises from green, it is simply not an applicable metaphor to say that mind creates body or mind arises from body. They are both *correlated results* from a single cause of possibility collapsing in consciousness.

This is a new form of psychophysical parallelism (see figure 4). In the old form, interaction dualism was avoided by postulating parallel functioning of the physical and the nonphysical worlds, but one could not answer the question, What maintains the parallelism of all the bodies running in parallel? Now we can say, consciousness does. Similarly, we have found the mediator of the interaction between the subtle body and the physical, if you insist on Cartesian interactional thinking. The mediator is consciousness.

41

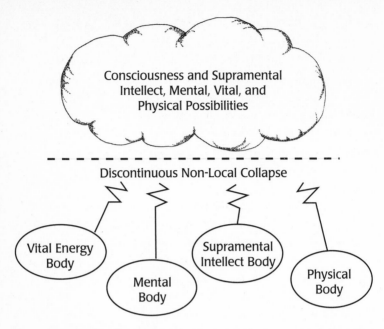

Fig. 4. Psychophysical parallelism. Consciousness contains four compartments of possibilities. With quantum collapse, the possibilities manifest as the material body, the vital body, the mind, and the supramental intellect.

Disease and Illness

More recently, medical researchers are finding it useful to distinguish between disease and illness. Disease is objective malfunctioning of the organism that can be diagnosed by machines, by suitable tests, about which experts can form a consensus. In contrast, illness is subjective, the subjective feeling of the malfunctioning. The materialist paradigm tries to explain disease but lacks the scope to explain the cause of the inner feeling or illness.

The quantum dynamics of all the bodies explains why part of our consciousness (the physical) is experienced externally, but part (the subtle body) is experienced as internal in our awareness. The explanation of this lies with the famous quantum uncertainty principle.

According to the uncertainty principle, firmly established by experiments to do with the movement of material objects, we cannot simultaneously determine both the position and the momentum (mass times velocity) of a quantum object with utmost accuracy. A way of seeing the validity of the uncertainty principle is to note the following. In the submicroscopic world, in contrast to the familiar macroworld, in an observation by us that requires some tiny signals like light to impinge upon what we are trying to observe, the impingement itself introduces uncontrollable tiny disturbances, hence, uncertainty. In other words, our observation affects quantum objects.

For the physical world, however, the world of extended bodies—*res extensa,* using Descartes' language—microbodies make up the macrobodies that then have heavier and heavier masses. For large masses, the quantum dynamics of matter is such that the objects expand as waves of possibility very slowly, so slowly that the effect of the uncertainty principle is hardly visible. So when your friend observes a chair in a certain position, and then you observe the same chair, your friend's observation affects the object's position so negligibly that you virtually observe the same chair in the same place.

So the two of you can compare your data and decide that since both of you are seeing the same thing, the thing must be independent of your observation, must be outside your awareness. That is, consensus data mesmerize us to conclude that this macro-world of matter is external to us. (But laser experiments show that objects like chairs do move by some imperceptible 10^{-16} centimeter between two observations.)

Now consider the subtle body. Here there are no extended bodies, and no micro-macro division. We have indivisibly an extension of worlds, infinite oceans of which the waves are experienced as individual events. But now the quantum uncertainty principle extends to all such waves; therefore one's observation always affects the object in the subtle body, so another cannot experience the identical object. Because of the lack of consensus, in this case we do not make the mistake that the objects are outside us. We experience them as private, and therefore we easily conclude that they must be internal.

So disease belongs to the physical body; it is external. Illness is internal—it is telling us about the malfunctioning of the simultaneously experienced correlated subtle body. If there were a one-to-one correspondence between disease and illness, there would be no problem; treating the disease would automatically treat the illness and vice versa. But empirically, there is no one-to-one correspondence: We can have a disease (early stages of cancer) but not feel ill. Or we can feel ill (the so-called psychosomatic pain), but there is no physical disease that can be found as the cause. Now do you see why we need an integrated medicine?

Representation-Making

What happens when consciousness simultaneously collapses the brain and the mind or the vital body and the physical body? First, the brain makes a map or representation of the mental meaning (see figure 5a). Or in the case of the vital-physical body, the physical makes representations of the vital morphogenetic fields, which themselves correspond to specific vital functions; these representations are the various organs in our physical body that carry out the particular vital function associated with the vital blueprint (see figure 5b).

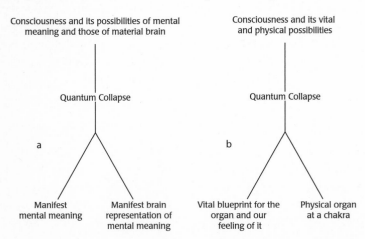

Fig. 5. (a) With the simultaneous correlated collapse of the mind and the brain, the brain makes a representation of the mental meaning that is experienced. (b) With the correlated collapse of the vital blueprint of an organ and the physical body, the latter makes a representation (the organ) of the former at a chakra.

In other words, the physical body acts like the hardware of a computer, making software representations of the vital morphogenetic fields becoming differentiated in the process called morphogenesis—form-making. Similarly, the brain acts as hardware and makes software of mental meaning.

What makes the physical body so suited as hardware? The same property of macroscopic fixity that makes the awareness of matter external in our experience. When you write on a chalkboard making representations of your thoughts, if the chalk marks were able to run away due to quantum uncertainty, it would not be convenient, would it? So the fixity of the physical world comes in handy for representation-making of the subtle.

The Levels of Disease and Healing

To proceed further, an idea from mathematics helps us. I am speaking of the idea of categories or logical types. A set is of a higher logical type than the members of a set. For example, think of prime numbers individually, and then think of all prime numbers. The latter is the set, and the individual prime numbers are the members of this set. This way of thinking is limited in mathematics because we cannot define a higher logical type than the set. The talk of the set of all sets gives us a paradox— Russell's paradox. There is no need to go into further details of mathematics here, but we can use the analogy with the mathematics of logical types to find an answer for the unification of medical paradigms.

I have already mentioned that quantum physics gives us a visionary window. If you look through this window, you have to turn your materialist worldview upside down. You see through the camouflage of the materiality of reality; you see all components of your experienced reality—physical sensations, vital feeling, mental thought, supramental intuition, and spiritual wholeness—as different levels by which consciousness experiences itself (Goswami 2000). These levels are nested (see figure 6).

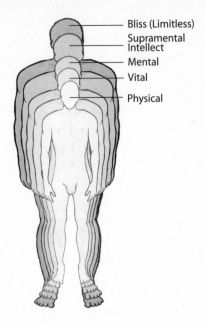

Bliss (Limitless)
Supramental
Intellect
Mental
Vital
Physical

Fig. 6. The five bodies of consciousness.

The physical is the grossest level, the vital a higher category, the mental a still higher category, and then comes the supramental (which healers do not usually invoke, categorizing it with the spiritual). If you are familiar with Platonic thinking, you can identify the supramental as the domain of what Plato called archetypes. In Jung's vision, this is the domain that we access by intuition. And finally, there is the spiritual, which is the whole—the ground of being, which we cannot experience in subject-object split awareness.

This picture of five "bodies" of consciousness is very old. It was discovered in India as part of the Vedanta literature (as in the story related earlier in this chapter) and also in the Judaic tradition as part of the Kabbala. Once you see the entire cosmology—world picture—in these terms, a veil lifts. You readily see that the different paradigms employed by different health practitioners are but a way of talking about different levels of disease and healing. And then you begin to see a way of integrating the different models to give you the ultimate holistic health paradigm, which is what Integral Medicine is about.

The physical level of disease seems easy to talk about: It is the body's normal physics and chemistry gone awry. The causes can be both external and internal. Examples of an external cause in the materialist paradigm are things like germs, viruses, and physical injuries.

Internal causes of a physical disease are more subtle, but an obvious one is a genetic defect. Deficiency of a gene or a combination of genes translates into the body's incapacity to make particular proteins for proper organ functioning—hence the disease.

But such an analysis as to the cause of the disease is not always possible. Take the case of cancer, for example. Both germ theory and gene deficiency have been considered as a cause, but not with much success. So the question, "What causes cancer?" is quite open to theories at the vital and mental levels.

What causes disease at the vital level? At the physical level, we have the physical body representations subject to the usual physics and chemistry; at the vital level we have body plans—the morphogenetic fields. An individual physical body is unique because of its structure. An individual's vital body is also unique but for a different reason—because of conditioning. Certain vital blueprints are used more than others, a group of propensities that then becomes a pattern of functional personality. Such an individual vital body: (1) may have certain imbalances built into it (internal cause); and (2) may acquire imbalances due to interactions with (a) the physical, (b) the vital, and (c) the mental environment (external cause).

Such an environment may consist of food, nature and animals, and other people. Note that the interactions of the vital with the physical and mental environment are indirect. Physical environment affects the physical body organs, but the latter are correlated with the vital body blueprints and so the effect propagates. Of course, consciousness makes the ultimate connection.

Similarly, the mental environment affects the correlated brain. The brain is connected to the various organs of the physical body through the nervous system and also through the newly discovered psychoneuroimmunological connections (see chapter 14). Finally, these organs are correlated with the vital body blueprints at the appropriate chakra. And consciousness again makes the connection.

These imbalances in the use of the vital body blueprints (the morphogenetic fields) then produce imbalances in the physical organ representations as well.

At the mental level, negative mental meaning may be attributed to external input occurring at all three levels:

1. At the physical level. For example, an injury causing mental anguish, "Why do these things always happen to me?"

2. The vital feeling level (the sight of a tiger producing fear and also fantasy fear).

3. At the mental level (words of insult).

The negative mental meaning affects the body through its representation in the brain, and subsequently through the brain's connection to the body via the nervous system and the psychoneuroimmunological molecules. Mental meaning affects the vital body blueprints directly at the crown chakra (top of the head) and indirectly through the physical body at the other chakras.

Additionally, the mind of an individual may also have built-in internal imbalances. Both internal and external imbalances of the mind are able to produce vital as well as physical imbalances.

Since the archetypal supramental level is not directly represented in the physical, there is no disease that can be said to be supramental in origin as such. But our lack of ongoing connection to the supramental and bliss bodies may manifest as ignorance that is the root cause of all suffering. The East Indian sage Patanjali (Taimni 1961) has said that ignorance gives rise to the ego, the ego develops likes and dislikes (a process I call mentalization of feeling), and these likes and dislikes eventually give rise to physical disease and fear of death.

Thus physical disease can be caused at all the levels, in all five bodies. The strict materialist assumes that all disease is caused at the physical level, and that is the most profound mistake of conventional medicine. But the alternative healing professionals make the same mistake if they attribute disease to any one level, as due to malfunction in any one body. In many cases, one must examine the cause of disease at more than one level.

Take the case of a physical injury. Materialists think this is a physical level problem. But the surgeons do their thing, and the wound does not heal. Now is the time to realize that the vital blueprints that assist the regeneration of the affected organ are not working properly. And this is the time to consult an acupuncturist.

The same holds for healing that must also be considered at more than one level. A disease comes with certain symptoms at the physical level, certain feelings of illness at the vital level, certain wrongness of meaning at the mental level, and a certain sense of separateness from the supramental and bliss levels. A complete healing is holistic healing—we should always try a multilevel approach, if compatible approaches for the different levels can be found.

Here is how it works. At the lowest level are conventional medicine and its materialist cures (of symptoms): drugs, surgery, and radiation. If the disease is entirely physical (which is seldom the case), then a material cure is the end of the story.

At the next level, the vital level, the disease has recognizable vital components as well as the obvious physical ones. If we treat only the vital components of the disease, as the Eastern practitioners of Ayurveda and traditional Chinese medicine do or as the homeopaths tend to do also, we have an exclusive paradigm. It is true that the treatment at the vital level is more fundamental and encompasses the physical level, but it takes time. So it is also true that in some cases when there is an urgency, the complementary use of the physical cure is clearly called for. The point is to focus on the compatibility of the two cures; then everything is moving in the right direction.

At the next level, the role of the mind is recognized; it is now mind-body disease and mind-body healing. Yes, at this level mind can be said to create the disease, but is it necessary to insist that mind alone heals, that it is all mind doing the healing at the mental level and that healing is percolating down to the physical?

Instead, why not continue compatible vital level and physical level healing as well?

In fact, one of the major points I make in this book is that mind-body healing is sometimes a misnomer. When mind creates the disease, sometimes the healing cannot be found at the level of the mind. One has to take a quantum leap to the supramental for healing. Of course, supramental healing does not exclude the mind; neither does it exclude the physical and the vital. A leap to the supramental fixes the wrongness of mental meaning; fixing the mental meaning fixes the vital feeling, signifying the healing of the morphogenetic programs so the latter can restore the biological functions of the organs at the physical level.

At the next level of spiritual healing, healing is recovery of wholeness (etymologically, *healing* and *whole* come from the same root) or what spiritual traditions call enlightenment. Some confusion arises here. If spiritual enlightenment is also the highest level of healing, why do supposedly enlightened people die of diseases such as cancer (so much so that Andrew Weil jokingly calls enlightenment an invitation to cancer). Why can't these enlightened people heal themselves?

It is a fact that two great enlightened mystics of fairly recent times, Ramakrishna and Ramana Maharshi, both died of cancer. But the confusion dissolves when you recognize that the discovery of wholeness heals the mind of the ego-separateness; the healing of the ego gets rid of vital imbalances due to emotional preferences, and no emotional preference means no fear of death at the physical level.

So there is nobody there either to suffer or be afraid of death due to the disease. Who, then, needs to heal it? In other words, the enlightened perspective may not make sense to perspectives from any of the lower levels!

Is Integral Medicine Science?

Practitioners of conventional medicine may still hesitate to embrace an Integral Medicine that integrates alternative medicine and conventional medicine. If medicine is generalized to involve nonphysical domains of reality (even conceding that they exist), would medicine still be a science? Science depends on consensus experimental data. Since by definition we cannot observe the non-physical with our physical instruments, how can we build a consensus science?

The answer to this kind of concern is not difficult. Our individualized nonphysical bodies, the vital and the mental, are not susceptible to direct physical measurements, true, but they have correlated effects in the physical that are available for laboratory experimentation. Moreover, we as conscious beings can directly feel, think, and intuit; these are our direct connections to the vital, the mental, and the supramental, respectively. If the doctrine of strong objectivity—namely, that science should be independent of subjects—is replaced by a doctrine of weak objectivity—namely, that science should be invariant from subject to subject—then medicine can be subjective and yet scientific.

The conventionalist may still hesitate: Suppose the so-called anomalous data of alternative medicine, mind-body healing, pain management by acupuncture, Ayurvedic *doshas*, homeopathic cures without physical medicine, spontaneous healing, prayer healing at a distance, all are real, but their nonphysical explanations are faulty and unnecessary. What makes you think that in some future time all these data will not find a perfectly material explanation? After all, we almost succeeded in showing that homeopathy is placebo healing (healing by sugar pills with doctor's blessing to enhance belief) and that acupuncture works through our nervous system (see later for details). The philosopher Karl Popper has called this attitude promissory materialism. Promissory materialism consists of vain promises that materialists make for solving a paradoxical problem or an anomaly in some future day with the help of additional materialist ideas that the future will bring.

For decades, promissory materialists have looked to replace the mind with nothing but the brain, but nobody has succeeded in building a computer that processes meaning. No biologist has succeeded in proving that the source of the programs of morphogenesis is contained within the genes or the cytoplasm (Lewontin

2000). Nobody can explain creativity without assuming quantum leaps to the supramental (Goswami 1999). And nobody has found a materialist explanation of the subject-object split of conscious awareness either. So these nonphysical bodies of consciousness are here to stay and we may as well use them to resolve the anomalies of conventional science and medicine.

The plight of the materialist reminds me of a story. A woman goes to a clothing store and wants to buy 50 yards of fabric for a wedding dress. The shopkeeper is surprised. "You don't need that much fabric, madam," he insists. "You don't understand," says our heroine, "my fiancé is a believer in promissory materialism. He likes to search, not find."

To summarize, the following comprise the modus operandi of Integral Medicine:

- Integral Medicine is based on a paradigm that most diseases occur simultaneously in more than one of the five bodies of consciousness—physical, vital, mental, supramental, and spiritual. However, the disease may originate in one level and spread to other levels.

- The goal of Integral Medicine is not to treat disease by targeting one level (the material) as in allopathy, but to target, as necessary, *all* the movements of *all* five bodies of consciousness as the field of healing.

- Specifically, both the mind and vital energies are accepted as places where disease can originate and healing may take place. Healing at a higher plane of consciousness heals the lower planes automatically, although it takes time.

- Naturally, the crude and invasive techniques of physical body medicine, at least in part, give way to subtler techniques.

- Illness and disease are clearly distinguished from each other.

- The idea of self-healing is accepted as part of the potency of downward causation of consciousness. Other-healing is accommodated as an example of nonlocality (see later).

• Physicians, therefore, once again, become cohealers with the patient (see chapter 6).

You can see that many of these ideas are already being practiced in alternative medicine schools such as in naturopathy. What is new here is quantum thinking, a conscious application of quantum principles to develop a thorough and workable system of healing. To their credit, many health practitioners have already intuited the importance of quantum thinking in medicine; they are already quantum doctors. This is the subject of the next chapter.

5

New Paradigm Thinking of a Few Contemporary Medical Practitioners

A paradigm is an umbrella of metaphysical premises and additional underlying assumptions and logical systems implicit or explicit in the day-to-day exploration of a group of scientists in a given field of human endeavor. Accordingly, conventional medicine has a working paradigm, with materialist metaphysics, classical physics, biochemistry, and molecular biology plus neo-Darwinism as its base.

Why do we need a paradigm shift in medicine? As noted by the philosopher Thomas Kuhn, who formulated the idea of paradigms and paradigm shifts, a paradigm is useful to its practitioners only until it begins to exhibit paradoxes that it cannot resolve and anomalous data that it cannot explain. Why is a paradigm shift in medicine needed? Because clinical trials are demonstrating the validity of alternative medicine practices, which constitute paradoxes for conventional medicine (see chapter 1). In addition, there are now definitive data for spontaneous healing, prayer healing at a distance, even placebo healing, which constitute anomalous data for the mainstream paradigm. Clearly a shift is called for to an integrative paradigm that can act as an umbrella for conventional and alternative medicine practices. I have already outlined this integrative paradigm—Integral Medicine—in the previous two chapters.

In this chapter, I want to delve into a little bit of history both in order to give credit where it belongs and to bring out additional paradoxes and anomalous data that conventional medicine cannot handle, but the new paradigm can (which will be the subject of the next chapter).

The Difficulties of Classical Physics Thinking

The truth is that most practitioners of medicine continue to think strictly according to classical physics even after a hundred years of the advent of the new quantum ideas. Classical physics gives us some invalid prejudices, the most blinding one being that there is an independent separate reality out there, and it is objective, that is, independent of consciousness. In medicine, this prejudice forces practitioners to ignore the causal role of the healer's and the patient's consciousness in healing, much evidence and even common sense notwithstanding.

Mary Baker Eddy (1906), who founded Christian Science, suffered from chronic illness for most of her life. In 1866, she suffered from an accidental injury that took her almost to death. Somehow she came out of this injury not only healed but with the insight that became the basis of Christian Science: Disease is unreal, a consciousness-created illusion, she said. And consciousness can cure it by de-structuring and restructuring its belief system.

Another prejudice that doesn't help is strict materialism, the idea that everything is made of matter and its correlates, energy and force fields. In this view, mind and consciousness are epiphenomena of matter. Elementary particles make conglomerates called atoms, atoms make molecules, molecules make the cells that make up the body including the brain, and the brain makes consciousness and the mind. If you take this view—the doctrine of upward causation, so named because all cause proceeds upwards from the lowest level of the elementary particles—then you must relegate consciousness to only an ornamental existence with no causal efficacy. Once again, this leaves no room for self-healing. You must also look at the mind as synonymous with the brain, leaving no room for meaning.

Yet many medical practitioners would at least privately admit that there is a role for meaning in healing, in what meaning the patient sees in the disease (Dossey 1989). But where does meaning come from? The brain looked upon as a classical computer cannot process meaning (see chapter 3). No, what processes meaning is the mind.

If matter is the only basis for things, there is also no room for such extraphysical objects as *chi* or *prana*, is there? Under the influence of material realism even practitioners of Eastern medicine have fallen prey to this materialistic emphasis and become defensive. For quite a while they have been looking for material explanations of their concepts, although the trend is now reversing.

With classical thinking, you either think of consciousness, mind, and the vital body as epiphenomena or you think of them as separate world entities, as dual, and then dualism—how can separate dual objects interact?—haunts you (Stapp 1995). So classical thinkers in medicine are forced to ignore well-substantiated data of mind-body healing and the well-appreciated success of traditional Chinese and Indian medicine and homeopathy, because the alternative is a philosophical blunder of attributing to the brain and the physical body the causal efficacy of consciousness working in conjunction with vital energy, and the mind.

There are data that squarely contradict other classical prejudices of medical practitioners. One such prejudice is continuity. Conventional medicine practitioners believe that the healing process is cause-driven and that these causes act in a continuous fashion. In this way the healing generated by these causes must also be continuous, gradual. Thus the prejudice of continuity translates as gradualism for the remission of disease. But there are now many well-publicized cases of spontaneous remission, including severe cases of cancer (Chopra 1989; Weil 1995; Schlitz and Lewis 1997), which are not gradual but sudden!

Another prejudice is the belief in locality—that all causes and effects must be local and must propagate via signals through space, taking a finite time. This one falls flat against the now famous documentation that prayer, even at a distance, even without a physical signal traveling to the patient, has the power to heal (Byrd 1988; see also Dossey 1989).

Quantum Thinking in Medicine

In 1982, the physician Larry Dossey wrote a book entitled *Space, Time, and Medicine.* I remember looking at the book; how could I miss it, being an avid reader of New Age books at the time? This was a time when even the now-famous Aspect's experiment definitively demonstrating quantum nonlocality, signal-less communication between correlated quantum objects, was not published yet. But Dossey was already talking about nonlocality in healing; he was urging medicine practitioners to give up the classical way of looking at space and time, with locality, and heed the message of quantum nonlocality, or action at a distance.

Six years later, the physician Randolph C. Byrd (1988) did his double-blind experiment on distant prayer healing. In this experiment, carried out at a San Francisco hospital, a group of patients were studied as to their healing rate, but only an arbitrarily chosen fraction of them were prayed for at a distance by a prayer group, without the knowledge of either the doctor or the patient. The result of the experiment is history: Those prayed for received the benefit of an enhanced healing rate! Prayer works for healing even at a distance. Nonlocality is important in healing! Can quantum physics be indeed important for medicine? Is the nonlocality exhibited in distant prayer healing an example of quantum nonlocality? (See chapter 6 for the answer.)

Then in 1989 came another seminal book on the possible application of quantum physics in healing. The book was called

Quantum Healing, and its author is the now famous Deepak Chopra, ex-endocrinologist, now Ayurvedic practitioner. Chopra was making the case for quantum thinking as the explanation of some cases of mind-body healing that seem like self-healing, patients healing themselves.

Conventional physicians are puzzled not only by self-healing, but also by mind-body healing in general because in their classical thinking either self and mind are brain, or they are dual entities, so it is dualism to entertain them as valid entities. Chopra suggested that perhaps the mind interacts with the body through a quantum mechanical body, and perhaps it is consciousness that helps mediate the interaction. Chopra was suggesting no less than downward causation by consciousness, quantum style. He was inspired to do this upon seeing discontinuous "quantum leaps" in self-healing.

The truth is that quite a few physicians seem to have overcome classical prejudices decades ago. I will mention one other—Andrew Weil. Even before Chopra, in 1983, Weil was already inviting his physician colleagues to look at quantum physics for guidance as to how to introduce consciousness back into the science of healing (Weil 1983). Not only that, Weil suggested that cases of spontaneous healing may very well be the result of a "flash of insight."

Weil cited the case of the patient designated as S. R., who was diagnosed with Hodgkin's disease (a cancer of the lymphatic system). Hodgkin's disease is known to progress in four stages; S. R. was already in stage 3. She was pregnant at the time and did not want to lose her baby, so she refused conventional treatment with radiation or chemotherapy and found a new doctor. Under his supervision, she had surgery, even radiation treatment, but the situation continued to worsen.

It just so happened that her physician was researching the application of LSD therapy for cancer patients. Under his guidance, she took a guided LSD trip during which the doctor encouraged her to go deep inside herself and communicate with the life in her womb. S. R. was able to do that when her physician asked her if she had the right to cut off the new life. It was then that S. R. had the sudden flash of insight that *she* had the *choice* to live or die. It took a while after this illumination, a lot of lifestyle changes to be sure, but she was cured. Incidentally, she also gave birth to a healthy child.

This is a clear incident demonstrating that we do have the capacity to choose our own reality, but to do so we need to be in a non-ordinary "illumined" state of consciousness.

Quantum collapse of the waves of possibility is fundamentally discontinuous. Although conditioning obscures this discontinuity or freedom of choice (Mitchell and Goswami 1992) in our normal functioning, it is available and its efficacy shows up in what we call the creative event—the flash of insight that Weil referred to in the case of S. R. So in the quantum view, spontaneous (and therefore discontinuous) events of healing can be seen as examples of creativity in healing.

Quantum collapse is also fundamentally nonlocal. Thus nonlocality of healing, as in prayer healing, finds straightforward explanation in quantum thinking.

In the next chapter, we delve even deeper into quantum physics than we have so far to understand the gifts of quantum physics to the science of life, health, healing, and death—the usual concerns of medicine.

6

More on Quantum Physics and Its Gifts to Medicine

If the title of this chapter is making you anxious, relax. The chapter is less on quantum physics and more on the gifts of quantum physics to new paradigm thinking in health and healing. From the preceding chapters it should be clear what three of these gifts are: downward causation, nonlocality, and discontinuity. There is another gift revealed by consideration of how quantum possibilities become actual events of our experience; it is called tangled hierarchy, a concept that I will leave a little mysterious until later.

Let's begin with a little history. What is a quantum? The word *quantum* etymologically comes from a Latin word meaning quantity, but the physicist Max Planck, who first introduced it in physics in a seminal work in 1900, had a slightly different meaning in mind. For Planck, and for quantum physics, the word quantum means a *discrete* quantity. For example, a quantum of light, called a photon, is a little discrete bundle of energy that cannot be broken down any further.

If the concept is still not clear, an example from everyday life may be of help. A cent or a penny is a discrete quantity of money; half a penny or half a cent does not exist.

Downward Causation

Quantum objects are waves of possibility. When we are not looking, they spread as water waves do when you throw a pebble in a pond. But a quantum wave spreads, not in space-time, but in the realm of possibility, a realm that Heisenberg called potentia. When we look, make a measurement, the wave of possibility collapses, what was spread out before (in possibility) becomes localized in actuality as a space-time event, what was many-faceted in potentia takes on one manifest facet (see figure 7).

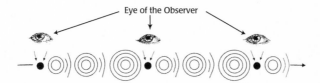

Fig. 7. Two phases of time development of a quantum object. A quantum object expands as a wave of possibility when nobody looks. This movement is continuous and is determined by quantum math. When we look, the possibility wave discontinuously collapses. This discontinuous movement is acausal and is not determinable by mathematics or algorithms.

Consider an example. Suppose we release an electron in a room. The electron's wave of possibility, if we are not looking at it, will spread in potentia. What this means is that the electron has the possibility of being all over the room in just a few moments. Each possibility, each possible position of the electron, comes with a probability forming a distribution (see figure 8). When we look, the wave collapses, the electron manifests in one of its possible places to be; an electron detector (for example, a Geiger counter) placed there ticks.

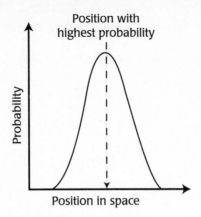

Fig. 8. A quantum probability distribution.

In the realm of possibility, the electron is not separate from us, from consciousness. It is a possibility of consciousness itself, a material possibility. When consciousness collapses the possibility wave by choosing one of the electron's possible facets, that facet becomes actuality. Simultaneously, the possibility wave of the electron detector also collapses, producing a tick; and the possibility wave of the observer's brain collapses, also registering the tick.

How the electron's wave, the detector's wave, or the brain's wave spreads in possibility, what facets these waves assume, is determined by upward causation, by the dynamics of elementary particle interactions. This part is calculable by quantum mathematics, at least in principle. The events of collapse of the waves of possibility are the results of conscious choice, downward causation. For this no mathematics exists, no algorithms. This choice of downward causation is free, unpredictable.

Discontinuity

Consider next the concept of discontinuity. The Danish physicist Niels Bohr gave us a picture of discontinuous movement that makes the concept crystal clear. Everybody knows that electrons go around the atomic nucleus in orbits, much like the planets going around the sun. That is continuous movement. But when an electron jumps from one atomic orbit to another, said Niels Bohr, the jump is discontinuous; the electron never goes through the intervening space. It disappears from one orbit and reappears in the other. Following Niels Bohr, we call this discontinuous movement a quantum leap (see figure 9).

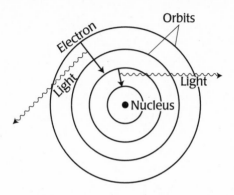

Fig. 9. A quantum leap as envisioned by Niels Bohr. According to Bohr, when electrons jump from one atomic orbit to another, they never go through the intervening space.

The mathematician John von Neumann (1955) further clarified the role of continuous and discontinuous motion in quantum physics. Quantum objects are described as superpositions of possible facets, or possibility waves. The possibility waves, von Neumann noted, develop in time in two clearly delineated ways. Between observations or measurements, their motion is continuous; they spread as waves in the domain of possibility, continuously, in causally traceable bits. But when we observe them in the process of quantum measurement, the possibility waves collapse discontinuously, from spread-out wave to localized particle, from a multifaceted object to one facet, all in one spontaneous, acausal step.

Nonlocality

Quantum nonlocality was introduced by none other than Albert Einstein, who in 1935 along with two collaborators, Boris Podolsky and Nathan Rosen, published a paper trying to discredit quantum physics. Einstein, Podolsky, and Rosen (1935) pointed out that a mere interaction binds two quantum objects into a nonlocal whole. The quantum collapse of the possibility wave of one part of such a system must instantly collapse the possibility wave of the rest. This is instantaneous action-at-a-distance. But nothing is allowed to occur instantly according to the theory of relativity. According to that theory, all signals that communicate action from one body to another must travel within the speed limit of the speed of light (300,000 km per second). But the three missed the message of quantum physics altogether.

Quantum collapse can be nonlocal and yet not violate the theory of relativity because it takes place outside of space-time. We must not picture quantum collapse like the collapse of a collapsible umbrella. The possibility waves of the two correlated parts of a system reside in the realm of potentia, outside space and time, where they are connected; on collapse the actual correlated events are discontinuously manifest in space-time. The quantum nonlocal connection that puzzled Einstein and colleagues lies outside of space and time; this quantum connection leads to a signal-less communication, and thus no violation of the theory of relativity is involved.

The Gift of Downward Causation

So how does all this relate to how we look at ourselves, especially in relation to our health and healing? Let's first consider downward causation.

When quantum physicists and quantum aficionados first had an inkling of the potency of downward causation back in the 1970s, many were elated. Soon the physicist Fred Alan Wolf coined the phrase, "we choose our own reality," and it became a New Age mantra. Many people started putting downward causation into practice, trying to manifest a Cadillac or some such thing via its help. And when that didn't work so well, they turned to manifesting parking spaces for their cars, Cadillac or not. But that did not work so well either.

Obviously, there are subtleties of downward causation that the 1970s' enthusiasts were missing. What are these subtleties?

One subtlety you already know: Who are we in relation to the world? Are we to apply downward causation to a world that is separate from us so that we don't have to be responsible for our action, or is the world us, and we have to accept responsibility along with our freedom of choice? In order to make sense of downward causation as a potent force in quantum physics, only the latter philosophy is acceptable—consciousness is the ground of all being. "We must supplement the Statue of Liberty on the East Coast by the Statue of Responsibility on the West Coast," said the philosopher Victor Frankl. So be it, says quantum physics.

So this much is clear. We cannot frivolously choose health over disease; we have to do it responsibly, such as following up our choice with suitable lifestyle changes. But can we even do it with what we ordinarily call choice, by wishing it? If we cannot manifest the car of our choice via wishful thinking, what guarantee do we have that wishing for health will manifest health, even if we are ready to promise responsible follow-up actions?

The question of who we really are is a subtle question, the mystics among us declare. We have to do much spiritual work, called *yoga* in Sanskrit (a word that means union or integration), to find out, they say. Fortunately, quantum physics—more precisely, considerations of quantum measurement—is giving definitive answers as to the nature of us, our consciousness. When you understand and integrate the lessons of quantum measurement theory in your life, you will be practicing yoga of a sort, no doubt. I call it quantum yoga—a scientific path to discover who we are.

In brief, considerations of quantum measurement tell us the following about the nature of our consciousness:

• Consciousness is the ground of all being.

• Matter, vital energies, mental meaning, and supramental archetypes are all quantum possibilities of consciousness.

• We choose, not in the ordinary state of consciousness that we call ego, but in a nonordinary state of consciousness that is variously known as unitive, nonlocal, or cosmic consciousness, a state in which we experience ourselves as one with everyone else.

• In an event of quantum collapse, consciousness splits itself into what we experience as subject-object awareness, subject experiencing an object as separate from it.

• Past experiences cloud our cosmic nature to an apparent individuality, the ego, via a process that can be called conditioning.

The first two you already know and perhaps have already incorporated in your being. You can see the importance of the third one immediately: We don't choose from our ordinary ego. So wishful positive thinking about our health will not necessarily give us health.

But how do we manifest our potential for downward causation then? The following pages of quantum measurement theory will give you strong hints. To make the discussion more interesting I will even throw in some related explanations of health and healing data. Are you ready for a little quantum yoga?

The Nonlocality of Consciousness

Consider a paradox first raised by the Nobel laureate physicist Eugene Wigner against the idea that quantum collapse consists of consciousness choosing actuality from quantum possibilities. That collapse is due to a conscious choice by an observer raises the specter of pandemonium in a case where there are two observers and two contradictory choices. To be concrete, consider the following scenario. Suppose you and your friend drive to a traffic light from two perpendicular directions. Let's say that the traffic light is a quantum traffic light with two possibilities: red and green. Now being busy Americans, you will both want to choose green, of course. If both of you get your choices, there is pandemonium. To avoid it, only one of you must be granted the power to choose. But on what criterion? Who gets to choose?

Wigner was puzzled because, to him, the only legitimate answer seemed to be a philosophy called solipsism—only you are real, and the rest of us, your friend included, are figments of your imagination. Then you are the chooser and there is no paradox.

Many people actually feel about the world in a solipsistic way. A Hollywood woman meets a long-lost friend, becomes excited, and invites her to "have a cup of coffee and catch up." But in her excitement, she just talks and talks and then suddenly becomes aware. "Oh, look at me, only talking about myself. Let's talk about you. What do you think of me?"

And yet we can appreciate Wigner's uneasiness because everyone feels solipsistic about everybody else. Fortunately, there is another solution that Wigner missed, which was independently discovered by three researchers (Bass 1971; Goswami 1989, 1993; Blood 1993, 2001): If it is always one consciousness choosing from behind our apparent individuality, the paradox disappears also. A unitive consciousness can choose objectively. So in a large number of such situations you and your friend each will get his or her wish half of the time; probabilistic anticipation holds. And yet this resolution leaves room for making a creative exception (as when there is a medical emergency) in any one particular case of looking.

So consciousness is one and universal, or as Erwin Schrödinger, one of the co-discoverers of quantum mathematics, put it, consciousness is a singular for which there is no plural. There are no two "consciousnesses"; our individuality is an illusory epiphenomenon of experience (discussed later).

So can we choose health over disease? Can we heal ourselves from a disease using the power of downward causation? Yes, we can, provided we develop the ability of transcending the ego and rising to unitive consciousness.

The late editor of the *Saturday Review,* Norman Cousins (1989), healed himself of a serious disease through laughter generated by funny movies and comic books. Although there is rumor that Cousins secretly used homeopathic medicine but was reluctant to admit it publicly, I have no doubt that his laughter therapy also substantially contributed to the healing. Laughter *is* when you are not taking yourself seriously. As the philosopher Gregory Bateson used to say, laughter is a half-step toward transcending the ego (more on transcending the ego later).

Quantum Nonlocality and Distant Healing

The physicist Alain Aspect and his collaborators (1982) verified quantum nonlocality in a laboratory experiment in which two correlated photons emitted simultaneously by an atom and moving away from each other were collapsed always in the same (polarization) state of actuality, although there were no signals between them. Yes, correlated quantum objects can influence one another at a distance without exchanging signals, by virtue of their quantum nonlocal connection.

Of course, experiments of this kind—and Aspect's is no exception—usually involve many decaying atoms and many pairs of correlated photons. Aspect's experiment reveals quantum nonlocality, but only after we compare and notice the correspondence of the (polarization) states of one photon at one detector at one place with that of the corresponding correlated photon at another detector at another place. But there is no correlation in the data collected in any one detector. These are entirely random. This is to be expected. Quantum objects are calculated as waves of possibility, and quantum mathematics enables us to calculate the probability associated with each possibility. In this way, quantum physics is probabilistic, and for a large number of events, randomness prevails. That is, the free choice that exists for individual events is always exerted so as to preserve the randomness in a large number of events.

Thus the quantum nonlocality that is revealed in Aspect's experiment is more like an event of what Carl Jung called synchronicity—meaningful coincidences attributable to a common cause. Two events take place at two different places. But you wouldn't see synchronicity—that there is meaningful coincidence—until you compare the two events.

Synchronicities are not uncommon in the healing literature. A doctor is enthusiastic about a new drug, a sample of which he received from a drug company. He administers the drug to his patient. The result is so impressive that he feels compelled to compare the effect of the drug with that of a placebo (sugar pill). But now the patient does not do so well. So when the doctor writes to the manufacturer for more samples of the drug, the manufacturer apologizes for sending him a placebo in the first instance by some mistake. The healing that took place is a clear case of placebo healing, but what prompted the manufacturer to make a mistake like that? Synchronicity or an Aspect kind of quantum nonlocality is a nice explanation.

In the previous chapter, I spoke of Randolph Byrd's data of distant healing, data pertaining to prayer group praying for double-blind patients at a distance and enhancing their healing rate as compared to control patients who were not prayed for. Can quantum nonlocality—Aspect style—be an explanation of this kind of data?

The answer is no. As I said earlier, since Aspect's data in any one place are random, and a meaningful message must involve a correlation between two subsequent events at the same place, there is no message in the data at any one detector location. Thus no transfer of message is possible through this kind of quantum nonlocal correlation between quantum objects.

In 1993, when my first book on quantum consciousness (Goswami 1993) was in press, I got a call from a University of Mexico neurophysiologist by the name of Jacobo Grinberg-Zylberbaum. Jacobo was doing an Aspect-type experiment to demonstrate nonlocal communication between human brains, but an aspect of the data was puzzling him, he said. At his invitation, I immediately went to Mexico to check his experimental setup. The experiment seemed quite legitimate. Here is what he was doing.

In the experiment of Grinberg-Zylberbaum and colleagues (1994), two subjects meditate for 20 minutes with the intention of direct (nonlocal) communication. After the 20 minutes, they continue the meditative intention but from two separate Faraday cages (electromagnetically impervious chambers), where each one is wired to an individual EEG machine. Then only one subject is shown a series of light flashes producing electrical activity in his brain, which is deciphered from its recording in his EEG as an evoked potential. Amazingly, his partner's EEG readings, when deciphered, show that the evoked potential, evoked by the light flashes, has been transferred to her brain as well, without any local connection. This experiment was subsequently replicated by the neuropsychiatrist Peter Fenwick in London.

The puzzling aspect of the experiment is that by looking at the transferred potential of one subject, you can conclude that light flashes have been administered to the correlated subject, even without checking his brain wave data. This is message transfer. What is happening?

The answer lies in the involvement of consciousness. In the case of correlated brains as in the experiment described, or in the case of correlated minds as in mental telepathy or distant healing, conscious intention is involved in establishing and maintaining correlation between subjects, the person praying and the person prayed for in distant healing. Ordinarily, as in the Aspect experiment, collapse breaks the correlation between correlated objects. Also the disparate events in any one place correspond to disparate objects. But in the Grinberg-Zylberbaum experiment (or in distant healing), consciousness maintains the correlation between the correlated brains (or minds), and the data in any one place always correspond to the same object—the brain (or mind) of the subject present there. So message transfer is allowed.

Don't think of quantum nonlocality as an esoteric concept. Let's elaborate the nonlocality of being alive—it is subtle. As modern human beings, we live more in our heads than in the body, but even so most of us would agree that there is a feel to being alive. The experience of this feel is unitary, not fragmented. We don't feel being alive in our big toe and our ears separately. There is an undeniable unity of experience here that gives you a direct sense of quantum nonlocality.

A related phenomenon is a big puzzle in neurophysiology: the binding problem. Now that we can take pictures of the brain while we mentate (Posner and Raichle 1994), there is no doubt that activity in several spatially separated brain areas accompanies our mental experiences. Neither is there any doubt that we have a unity of experience. So the neurophysiologist worries, How do the disparate processes in different brain areas bind together to give us the unitary experience? It is a clear case of quantum nonlocality.

Tangled Hierarchy: Dependent Co-arising of Subject and Object

One of the surprising things in the event of quantum collapse is that when you look, not only does an object appear in consciousness but also a subject appears looking at the object. Quantum collapse produces the awareness of a subject-object split—the experience of a subject looking at an object. This can be understood by examining the role of the brain in making a conscious observation. No experimenter, no human observer, has ever performed a quantum measurement, a quantum collapse, without a brain! According to quantum rules, before measurement, before collapse, not only the object/stimulus but also the observer's brain itself, the brain that is taking in the stimulus, must be represented by a wave of possibilities. There is circularity here: Without the brain, there is no collapse and no awareness, no subject and no agent of downward causation; but without collapse, there is no actualized brain. The resolution of the circularity is dependent co-arising.

In the event of a quantum measurement, the collapsing subject and collapsed objects, including the brain, arise simultaneously, codependently. The experiencing subject and the experienced objects cocreate one another. The subject sees the object as separate from it—this is called self-reference. But it is only appearance; the truth is that consciousness creates both subject and object. Both the brain and the object are collapsed in the same event, but we never experience the brain as an object. Instead, consciousness identifies with the brain that is then experienced as the subject of the experience.

The dynamics of dependent co-arising can be understood using the idea of a tangled hierarchy (Hofstadter 1980). You know simple hierarchy; it occurs when one level of a hierarchy causally controls the other(s), but not the other way around. Go back and look at figure 1, which depicts a simple hierarchy. To understand a tangled hierarchy, examine the liar's paradox: *I am a liar.* It is a tangled hierarchy because the predicate qualifies the subject, but the subject qualifies the predicate also. If I am a liar, I am telling the truth, then I am lying, and so on, ad infinitum. The tangle can be seen (and also resolved) only by "jumping out of the system." We cannot see it if we identify with the system. Instead, we get stuck and think of ourselves as separate from the rest of the world.

So quantum measurement involving the brain is tangled hierarchy. The reward is that we gain the capacity for self-reference, the ability to see ourselves as a "self" experiencing the world as separate from us. The downside is that we don't realize that our separateness is illusory, arising from a tangled hierarchy in quantum measurement, quantum collapse.

You may have seen the Escher picture of "Drawing Hands." In that picture, the left hand draws the right hand, and the right hand draws the left, producing a tangled hierarchy. Again, if you identify with the picture, you can get caught in the infinite oscillation of the tangled hierarchy. But is the right hand really drawing the left? Is the left hand really drawing the right? No, from behind the scene, Escher is drawing them both.

So is the subject collapsing the object? Is (are) the object(s) creating the subject? Neither. From behind the scene, consciousness, through the illusion of a tangled hierarchy in quantum measurement, is becoming both, the subject and the object(s).

Tangled hierarchy and self-reference are important for us to understand. A lot of mental stress develops in our lives because we grow up in a dysfunctional family. A family is dysfunctional when it fails to act as a unit, when it lacks a "self" identity. Such a family self-identity arises only when there is tangled hierarchy in the relationships between the family members. The same thing is true for couples, and more to the point of the present discussion, true for the doctor-patient relationship.

I submit that tangled hierarchy, no doubt because of its conceptual difficulty, remains one of most unappreciated principles of nature, but if you are interested in healing, you must heed tangled hierarchy. Just notice one of the undeniable sources of popularity of alternative medicine over conventional medicine. A conventional medicine doctor treats you in a simple hierarchical manner. He or she dictates, you listen. But most alternative healers practice tangled hierarchy in their relationship to their patients! They dictate *and also* listen. You and your healer then become a self-referential unit. Such a self-referential unit has value. It enables you (together) occasionally to take creative quantum leaps of healing (more on this in chapter 16).

The Distinction between Conscious and Unconscious

The idea that our subject-object split awareness arises from a quantum collapse enables us to understand the enigmatic concept of the unconscious that Freud introduced. We have seen above that awareness arises with quantum collapse. The unconscious is operational when awareness is not, when there is no quantum collapse. The unconscious is a misnomer in a worldview based on the primacy of consciousness, because consciousness is always present. What Freud meant was "unaware," absence of awareness.

The concept of the unconscious is important for the subject of health and healing in connection with psychosomatic disease. We suppress the memories of certain traumatic experiences so deep that consciousness seldom collapses them, delegating them to what is called unconscious processing. The memories of these experiences are processed by producing somatic effects of disease, but we are not aware of them, because we never collapse these memories in our conscious thoughts (more on this in chapter 15).

Somebody is uneasy about feelings and suppresses them, relegating them to the unconscious, creating blocks in the flow of vital energy. These blocks eventually lead to malfunctioning organs when we have a conscious experience of disease. But we are not aware of the energy blocks that are responsible for the disease (more on this later).

Unconscious Processing and Creativity

Creativity is an undeniable component of biological beings. We can see the similarity of the two modes of movement of the quantum object—continuity and discontinuity—in two important components of the creative process. It is well known that the creative process consists of four distinct stages (Wallas 1926): preparation, unconscious processing, insight, and manifestation. The first one and the last one are obvious—preparation is reading up and getting acquainted with what is already known, and manifestation is capitalizing on the new idea, obtained as insight, by developing a product; these stages are both done more or less in a continuous fashion. But the middle two processes are more mysterious. They are the analogs of the two stages of quantum dynamics.

As discussed previously, unconscious processing refers to processing during which we are conscious but not aware; we process the possibilities but remain inseparate with them. In creativity, unconscious processing is believed to account for the proliferation of ambiguity of thought. It is the analog of the spreading of the quantum possibility wave between measurements (see figure 7). Creative insight, on the other hand, is found to be sudden and discontinuous. It is the analog of the quantum leap, a discontinuous leap of thought without going through the intermediate steps. Unconscious processing produces a multitude of possibilities; insight is the collapse of one of these possibilities (the new one of value) to actuality.

Thus, once we permit quantum thinking in our science of us, we make room for both continuous and discontinuous processes; we make room for creativity.

Quantum healing, a concept introduced by the physician Deepak Chopra (see chapter 5), is the result of a creative quantum leap. This will be further discussed later in this chapter and in chapter 16.

Measurement, Memory, and Conditioning

What is the nature of the subject/self of self-reference that arises from tangled-hierarchical quantum measurements? Consciousness identifies with the brain that becomes the subject of the resultant subject-object split. This identity I will call the quantum self. In this identity, the self is universal (that is, it has no personality), and the choice from possibility to actuality is free and potentially creative.

Much confusion arises in relating to this picture because this is not the self we regularly experience in waking awareness. How do we get from the universal, unitive, quantum self-identity to the local and personal ego-identity? The answer, in a nutshell, is conditioning.

Experiences brought about by quantum measurements in the brain produce memory; a repeated stimulus is usually experienced, reflected in the mirror of past memory, through secondary-awareness processes (in contrast, the first collapse event in response to a stimulus is called a primary-awareness event).

This reflection in the mirror of memory (see figure 10) reinforces the probabilities of the subsequent collapse in favor of the conditioned response (Mitchell and Goswami 1992). I will call this quantum memory as opposed to ordinary content memory, which requires a macro-body. Over time, all our responses to learned stimuli comprise a habit pattern. The quantum self-identity which is natural for a young child gradually gives way to an identity with a particular history and habit patterns, an identity that we call the ego (Goswami 1993).

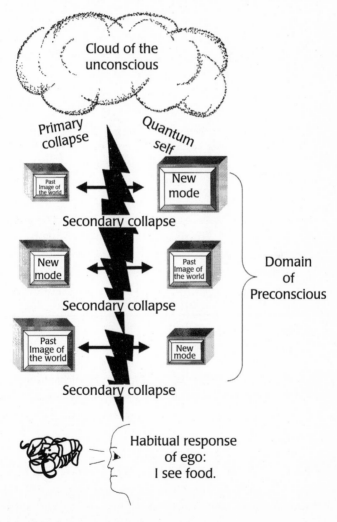

Fig. 10. Processing via reflection in the mirror of memory produces the conditioned ego.

A quick aside: Since our mind is correlated with the brain, as we develop a brain-individuality via quantum memory, we also develop quantum memory and habit patterns of the mind, an individual mind. The same thing can be said of our physical body–vital body duo; experiences produce an individual vital body with individual vital propensities.

These mental and vital propensities are what Easterners call karma, and that plays a crucial role in the scientific theory of reincarnation (Goswami 2001). When an individual's physical body dies, what survive are the vital and mental bodies with their vital and mental karma. This karma is recycled to the next incarnation.

There is evidence in favor of such a scenario of our self—physical, vital, and mental. This theory leads to the following model for ego development. As we grow up, we are creative in our quantum self-identity, continually discovering new contexts of living, of being human. As we discover a new context, we also explore the secondary contexts available to us by combining new ones with old learned contexts—we adapt and assimilate; this is a stage of homeostatic adaptation. This model of alternative creative spurts and homeostatic adaptation for ego development is substantially the same as that arrived at by the psychologist Jean Piaget (1977) as a result of his long series of experiments with children (see Goswami 1999, for further details).

That there is a time lag of half a second (Libet et al. 1979) between the objective time of the arrival of a stimulus to the brain and the (subjective) time of our usual waking awareness of it gives further credence to the scenario. And although the quantum self-experience of primary awareness is usually relegated to what psychologists call the preconscious, we do penetrate the preconscious when we are creative.

Creativity researchers call this entering the preconscious the *flow experience* (Csikszentmihalyi 1990). When we are spontaneous and alive in physical activity such as a dance (Leonard 1990), when we feel one with the universe in a sudden moment of spiritual surge, when we are in meditative awareness, we are in flow. The experiencer tends to merge with the experience. Harvard neurophysiologist Dan Brown (1977) established that meditation reduces the reaction time of secondary processing.

There are reports of such flow experiences from cancer patients undergoing spontaneous remission. The physician Richard Moss (1981) tells us an anecdote that illustrates this idea. Moss runs workshops featuring a lot of bodywork and in his early days, in the 1980s, his workshops were famous for healing. So a terminal cancer patient, a woman whose cancer had spread all over her body, came to one of Moss's workshops. Although she must have come for healing, initially she was quite unwilling to participate in the rigor of the practices of the workshop.

Moss kept prodding her, sometimes to a degree that could be called insulting. After this went on for a while, in one session of active dancing, the cancer patient was so angry at Moss's prodding that she overcame her diseased body's lethargy and danced. And she really danced! Next morning she felt much better and tests showed that her cancer was gone.

I submit that while dancing with abandon, this patient forgot herself, she transcended the ego, went into her preconscious, and entered the state of flow. The dancer became the dance. She became available for the creativity of the quantum self. Eventually she took the quantum leap! And her cancer found the overnight cure of quantum healing.

Quantum Gifts to Medicine

This whole book is about the gifts of quantum physics to medicine. Here I will count the many ways quantum gifts are bestowed on health and healing:

1. Quantum physics allows us to integrate all the disparate philosophies of the various schools of medicine. This I have already demonstrated (see chapter 3).

2. Quantum thinking enables us to develop a useful taxonomy of disease and healing. This classification we have already developed (see chapter 4).

3. Quantum physics shows clearly that we can choose between disease and healing. We can exercise this choice once we get the hang of quantum leaping to unity consciousness, the quantum self.

4. Quantum physics enables us to understand anomalous phenomena of medicine such as spontaneous healing (as instances of quantum creativity), distant prayer healing (as instances of quantum nonlocality), and self-healing and spiritual healing (as downward causation with pure intention) (see chapters 16 and 17).

5. Quantum physics even clarifies the role of allopathic medicine in integral healing (see chapter 7).

6. Quantum physics gives clear guidelines for the doctor-patient relationship (tangled hierarchy) (see also chapter 16).

7. Quantum physics clarifies and explains many hitherto mysterious facets of Eastern medicine (both Chinese and Indian), chakra medicine, homeopathy, and mind-body medicine (see parts 2 and 3).

7

The Place of Allopathy in Integral Medicine

In 1996, thanks to Marilyn Schlitz of the Institute of Noetic Sciences, who was on the inviting committee, I had the good fortune of being an invited speaker at the Tucson biannual conference on consciousness. Not as a plenary speaker, mind you, but that was okay; after all, this conference focuses on material aspects of consciousness, mainly neurophysiology and behavior, and pays only lip service to the mainliners of consciousness research. I was part of their lip service, but still I considered it an honor and gave a highly spirited talk (judging from inside it).

What happened next is the reason I am telling you this story. When I sat down after the talk, my heart was highly disturbed, I could even call it pain. Although I recovered in half an hour, as soon as I was back home, I went to my doctor. He saw angina in my description and sent me to a heart specialist, who gave me an angiogram and diagnosed many blocked arteries. When I was asked if I wanted bypass surgery, considering that there is only a one in 1,200 chance of failure, I did not hesitate a moment and said yes. Needless to say, the procedure was successful.

Here, then, is what I am trying to say. Allopathic medicine is useful and wonderful when it is useful. The point of this book is not to slight allopathy, but only to complement it as necessary.

This is also not a place to summarize the basics of allopathy. First, almost every educated person knows the basics anyway. Second, it is boring. Third, we don't need it. In a few situations where we will need to discuss some basic anatomy and physiology, we will do it in that context.

Instead, I want to make some general comments about allopathy from an outsider's point of view with the hope that not only you, the lay reader, but also the occasional professional will engage with the comments and find them useful.

To set the context, let me reiterate that the healing methods of allopathy are strictly material: surgery, radiation and other physical therapy, and most important, drugs. (We may add to this list gene therapy and behavioral modification in accordance with the social theory of disease [see later], but these are also materially based procedures.) Sometimes we hear complaints that allopathic medicine does not have a theoretical basis for healing since it only attends to disease. Medicine, in allopathy, is disease control and management. But *that* is the theory.

Allopathic medicine looks upon the human body as a car mechanic looks at an automobile. Materialist biology makes no distinction between the living and the nonliving, and allopathic medicine is based on that biology. Can we define the health of a car? You bet. When the car is functioning well! When the car gives symptoms of dysfunction, then is when we seek the advice of a car mechanic. Why should it be any different for the human body, when it is also a machine?

We may not agree with this philosophy, but allopathic medicine is consistent, and that is a strength. Many people criticize allopathy because it is reductionistic. To them, medicine should be "holistic."

What does "holistic" mean? The word "holism" was originally coined by Jan Christian Smuts, who among other things, served as the prime minister of South Africa for a time. His idea of holism is the whole is greater than the sum of its parts. The idea of holistic medicine is to introduce nonmechanical concepts into the healing equation without breaking from the basic materialist doctrine, that everything is matter.

Holists who are followers of Smuts say that there are apparent nonmaterial aspects of ourselves—any complex system should have them—that emerge from the material and are not reducible to the material aspects of us. However, in truth, nature does not give any indication of this kind of holism when it makes conglomerates from simple stuff. When atoms aggregate to make molecules, no "greater than the parts" epiphenomenon emerges that cannot be explained by the interaction of the parts.

Well, then, materialist medicine better be reductionistic, because reductionism is the way that the material universe seems to work. Is holism then out as a metaphysics for medicine? The defender of Smuts's kind of holism may say that acupuncture is an example of holistic medicine. The acupuncturist puts the needles in the foot to treat the heart! But these holists have a misunderstanding of the holism of acupuncture. The acupuncturist is not primarily treating the physical body through his treatment. His primary target is the vital body; the physical effect is secondary (see chapter 10).

Gradually, a new kind of holism is coming into vogue. In this usage, holism means to incorporate the *whole person*—body, mind, energy body (vital body), even soul (meaning supramental) and spirit (meaning what we call the bliss body). This book is only one of the latest to incorporate this kind of holistic thinking within a framework that follows naturally from quantum physics.

Allopathic medicine is like classical physics. In the realm in which it is applicable, classical physics is useful. The same is true of allopathy. It's just that the applicability of allopathy is limited by its metaphysics. So what is the domain of allopathy? When is it useful?

Medical Emergency

The Buddha, when asked about one of what he called "the 14 questions," used to say that if one has been hit with a poisoned arrow, is a metaphysical discussion useful or isn't it better that we take out the arrow first? This is the first thing. The physical body, our gross body, makes representations of the subtle. But this representation-making capacity is vital; without it, the subtle bodies will no longer be able to correlate with the physical, and death will result. So we must attend to fixing the physical before we worry about our subtle bodies. And if that means any of the allopathic procedures—surgery, radiation, or drugs—so be it.

In my own case, I did not hesitate with bypass surgery for this reason. I knew the alternative; I knew that angina can be reversed, thanks to Dean Ornish's research (which showed the efficacy of diet, exercise, and meditation to this effect; see Ornish 1992), and I also knew that I had the stamina to change my lifestyle to reverse the problem. The only thing I did not have was time.

For emergency situations, the gross methods of allopathy are often needed, side effects notwithstanding. Before allopathy, medicine was natural medicine even at the physical body level. For example, medicinal plants as such were employed for treatment of disease. Modern allopathy changed that. It was found using clinical methods that if we extract the chemical relevant to the treatment of the disease of the appropriate organ, the efficiency of the treatment improves, sometimes enormously. This is reductionistic, no doubt, but it is quite appropriate for the material level of treatment that allopathy professes.

So whenever time is a problem, allopathy gets the nod and its defects have to be tolerated. If you have pneumonia, acute strep throat, or basillary dysentery, go allopathic; use an antibiotic although it is going to mess up your intestinal flora. In those cases, allopathy has to be the primary care, although secondarily, we can and we should also pay attention to the subtle bodies. There are ways to do so, or we should discover ways. For how long should allopathy dominate our treatment? Only until the emergency is over. Then is the time for looking for alternatives to allopathy.

The Correspondence of the Old and the New Paradigms

When the philosopher Thomas Kuhn introduced the idea of paradigm shift (see chapter 5), he was very clear about one thing.

Since scientific paradigms are based on verification, experimental data, the old paradigm is valid as far as it goes, as far as it is applicable. The new paradigm does not make the old one wrong, but only points out its limit and extends science to a new horizon.

This idea is so important in considerations of a paradigm shift that it is made into a solid scientific principle called the correspondence principle: In the limited domain in which the old paradigm is valid, the new paradigm smoothly corresponds to the old paradigm. The correspondence principle enables us to go on using the old paradigm in its old domain of validity. No paradigm battle is necessary.

Let's recognize medical emergency as the domain of validity of allopathic medicine. In this limit, as seen above, the healing of the subtle bodies must give way to the healing of the physical body without which the subtle bodies cannot function. Thus we can define a clear correspondence principle for the paradigm shift from an exclusively allopathic medicine to an integral holistic medicine: With the limits of medical emergency, the new integral holistic medicine can be replaced by allopathic medicine for all practical purposes.

Actually, there are a few natural medicine remedies quite suited for emergency as well. For example, the homeopathic medicine *Arnica* works better than allopathic drugs as first aid for shocks, burns, and other traumas. If you are worried that this violates the correspondence principle, don't be. An occasional violation of the principle is allowed. This is why it is called a principle, not a law.

Alternative or Complementary Medicine?

I now can tell you what a true alternative or complementary medicine must do: It must take healing beyond the material dimension of ourselves. When an alternative or complementary medicine becomes comprehensive or integral, then it should be able to take care of all five of our bodies simultaneously with appropriateness. This is also a true holistic medicine. And here is the problem with allopathy and its reductionism: The reductionist methods don't match well with taking care of *all* the bodies at the *same* time.

For example, take the case of narcotics for pain relief. They are effective, and almost everyone uses them in cases of emergency. But you cannot use narcotics if you also want to take care of your disease at the mind level, as that requires your facility of awareness to be intact. In general, allopathic drugs affect the physical body quite globally and not locally where the effect is intended. Invariably they interfere with the physical body's other functions. For example, they interfere with the capacity of the physical body to make new representations of the vital and the mental. Since reaping the benefits of the healing of the vital and the mental involves the physical body's representation-making capacity, the impairment of this capacity is acceptable only in emergency situations, not at any other time.

I am not proposing any hard-and-fast rule here, just common sense. There will be exceptions to the above rule. Aspirin may be one; cholesterol-reducing drugs may be another; perhaps even Viagra is an exception.

But in general, allopathic drugs should be avoided whenever you are treating or engaging with one or more of your other bodies, whenever you are using integral, holistic medicine. So then it is better to use natural plants and herbs instead of the allopathic extraction. Remember, time is not a factor in Integral Medicine, which we seldom apply in emergency.

When should we apply Integral Medicine? Later in this book, we will see that alternative medicine systems are primarily designed for our subtle bodies, and only secondarily for the physical body. The idea is to cure the vital body imbalance or mental body imbalance that makes us ill. But it takes time. The subtler the body, the more time it takes to treat the imbalance. It takes more time to treat a mental imbalance than a vital one. So depending on the time factor, we choose.

For a cold, which is going to last a few days only, but is not an emergency either, choose vital body medicine—Ayurveda or Chinese medicine—over mind-body techniques such as meditation or biofeedback, even though the root cause of your cold may be in your mind. But if it is heart disease, after the emergency is over, mind-body medicine is preferable because since you have bought the time (and at what cost!), you may as well go to the root of your problem, which is the mind. Treating the mental will automatically address also the vital and the physical. That is how it works.

Other Uses of Allopathy

So should allopathy become only an emergency medicine in the approach of Integral Medicine? Not entirely. There are two more important uses of allopathic medicine.

One use of allopathy that should stay is allopathy's version of preventive medicine, practically its only preventive use. This is a little surprising, this nonemphasis on prevention in allopathy. For our cars, we certainly don't preclude preventive care. Anyway, the one prevention technique that every allopath heartily approves is vaccination. But some homeopaths take a very different view.

It is important to hear out the homeopaths on the concept of developing immunity against disease through vaccination. The homeopathic physician Richard Moskowitz argues on the basis of his own clinical experience that immunizations may prevent acute disease, but they make the body vulnerable to chronic disease later in life. This is because vaccines make the immune system weaker (see Leviton 2000, for further discussion).

Perhaps we should then heed the lesson of the correspondence principle (see earlier discussion) here and reserve immunization through vaccination to emergency epidemic situations only.

The other area in which allopathic medicine will continue to be useful is the area of diagnostic techniques, limited as they are to pointing out the disease of the physical body alone. By and large, the diagnostic techniques of alternative medicine, by necessity, require a lot of intuition, and even then, the diagnosis is never a sure thing. It is important to be sure about at least the gross body wrongness when we are treating disease.

The Rise and Fall of Allopathy

I enjoyed reading a book, *The Rise and Fall of Modern Medicine,* by the physician James Le Fanu (2000). The author gives us a history of modern medicine through its rise with the discovery of antibiotics and the success of open-heart surgery and organ transplants and its fall when it became clear that there may be no more miracle drugs on the horizon or new breakthrough techniques of surgery.

What about the frontiers of allopathy—gene therapy, for example? According to Le Fanu, gene therapy may not be the panacea that some researchers claim.

What is gene therapy? It consists of the correction of genetic defects (that are known to give rise to disease) by replacing the defective genes with normal genes. Can such a replacement be carried out? Researchers had the brilliant idea of neutralizing a virus (by removing its harmful genes), injecting the virus with the normal gene, and then introducing the modified virus to invade the cells containing the defective genes. Unfortunately, the procedure has not lived up to its promise.

According to Le Fanu, the other frontier of allopathy today is based on the so-called social theory—the epidemiological idea that diseases such as cancer and heart disease are caused by unhealthy lifestyles and environmental pollution. So treat the disease by (a) making lifestyle changes and (b) reducing environmental hazards. That this can be considered as a frontier of the materialist-reductionistic system of allopathy is itself interesting. What constitutes a lifestyle change? To the allopath, it is healthy diet and exercise. But how about mental belief systems that attribute meaning to the environmental stimuli? Could changing those belief systems lead to the cure of "social" disease? What constitutes nutritious food? Should we include vital energy considerations in our discussion of nutrition?

Thus the allopath's acceptance of the social theory of disease easily leads to questions that can open allopathy up to accepting techniques of Integral Medicine such as mind-body healing and Eastern systems such as Ayurveda and Chinese medicine.

Le Fanu does not envision this. He wants answers based more on traditional biology, such as germ theory. He finds light at the end of the tunnel in the so-called bacterial explanation of peptic ulcers—the idea that ulcers are caused by the bacterium *Helicobacter*. But if you treat the bacterium with antibiotics, the cure is often only temporary, suggesting that the bacterium is not the cause but only an associative factor in severe peptic ulcers.

No, I don't think biological answers are going to come for the diseases that allopathy cannot treat; no material answers are going to come. The fact is that our being is more complex and extends beyond matter. Matter is hardware and is important, but those subtler aspects of us that the hardware makes representations of are equally important and must be taken into account in a proper science of healing. Allopaths must come to terms with reality.

When I was a student, one of my teachers gave me advice that made an impression. You look at a problem with a preference for the kind of answer you like, that you think should be the answer. You try your best to match the question and the answer based on your prejudice, because *what else can it be?* But after you have tried and failed so many times, then you can give up. *What the hell!* You are ready to consider alternative answers. By and large, I have found this very useful advice. If your belief in allopathy is paramount, I suspect this may also be good advice for you.

Biology within Consciousness

What conventionalists in medicine should note is that biology, the science they accept as the paradigmatic basis of medicine, is in dire need of a paradigm shift. The hints have been accumulating for quite some time.

Let's start with the problem of consciousness. Neurophysiologists try to apply their reductionist methodology to understand consciousness as the product of brain processes, neuronal interactions. But as the philosopher David Chalmers has pointed out, how can this approach succeed? A reductionist approach can only succeed in making a model for an object in terms of simpler objects, but consciousness is not just an object, it is also a subject.

If biology cannot explain consciousness, it is time to consider whether a metaphysical basis of the primacy of consciousness, the one that quantum physics is giving us (see chapters 1–6), can explain the various unexplained phenomena of biology.

For a proper biology within consciousness, we must assume that a single living cell is already wired for self-referential quantum measurement. Suppose the quantum measurement for a living cell is also a tangled hierarchy, similar to the case of the brain. As consciousness collapses the states of the cell, it identifies with the cell self-referentially, an identity that we call life, as distinct from the environment.

In its fundamental nature, this identity is an identity with all of life, since all life originates from that first living cell. Following James Lovelock (1982), I call this identity "Gaia consciousness." This fundamental identity then propagates further down in an alternating play of creativity and conditioning, much like our ego development (see chapter 6).

I hope you notice that in the present view, the separateness of life and environment is only an appearance arising from the tangled hierarchy of quantum measurement. The environment, nature, is not really our enemy—the giver of disease—and we are not its victims, as allopaths often tell us. The environment is us. Its separateness from us is illusory play. It can even be argued, as the author Richard Leviton (2000) does, that disease is part of the covenant of our life; it can be taken as our teacher (more on this in chapter 17).

Next, consider evolution. There is evidence that biological evolution is an interplay of conditioned action (at best, previously learned contexts are combined to solve a problem with changes in the environment) and creativity (a new context is discovered that gives us new problems as well as a new solution). During periods of species homeostasis in which previously learned contexts of living are combined to produce a situationally new context of living (a process called adaptation), continuous Darwinian evolution is at work. There is plenty of fossil evidence for this.

But when there is a quantum leap to a truly new context of living, the creativity of rapid quantum evolution reigns (Goswami 1997). Because of its rapidity, this phase does not leave any fossil remnants; there isn't time (Eldredge and Gould 1972).

So life begins with the self-referential collapse event of the first single cell, but what is the meaning of the evolution of life in this view? Life evolves toward more complexity, toward making better and better (more suited to express the archetypal themes of consciousness) physical representations of the vital body blueprints of form-making to perform better the archetypal functions of living. Eventually, when the brain is made, the mind can be mapped. Before the evolution of the brain, mind could only be mapped indirectly, through the intermediary of vital body representations of the mind.

Once a physical body representation is made, the next time the representation is collapsed to perform a function, the correlated vital body movement collapses also, which we experience as a feeling. Similarly, the collapse of a brain state that is a representation of a mental meaning automatically brings on the collapse of that mental state of meaning, which we experience as a thought.

Notice, however, that at this stage of the evolution of life, there is no physical hardware that can make representation of the supramental intellect. So representations of the supramental in the physical are always indirect, made through the intermediary of the mental and the vital bodies, and therefore imperfect.

Note also that the living world is an identity of consciousness, but the identity occurs at many different levels; it is not a single identity. First, there is the identity with the whole of evolving life on Earth, Gaia consciousness. Second, there is species identity, an identity with a particular class of forms and adapted genetic habits. Third is the identity of the individual organism. But the identities do not stop there.

Each living cell of a multicellular body has a self-identity; consciousness identifies with each of them as it carries out its individual conditioned functions. Any conglomerate of cells such as an organ that involves self-referential quantum measurement at the conglomerate level of functioning also has self-identity.

For an organism with an integrating brain, consciousness identifies with this particular group of cells in such a spectacular way that it obscures much of the other body identities. This brain dominance of our body functions has taken our attention, particularly the attention of researchers engaged in a scientific study, away from the body organs and the feelings of vital energy that arise in the body organs at the various chakras.

Richard Leviton (2000), in a critique of the current organ transplant boom, writes:

What about the personality and energy residue of the transplanted organs? Does my liver have anything to do with who I am?

If you ask a doctor of Chinese medicine, the answer will be yes: my liver carries a definite signature of my energy, my style of Qi.

Biology within consciousness agrees with Leviton and the Chinese physician: Our livers do have something to say about who we are. The consciousness connection of our organs precludes the kind of thinking that has led to the organ transplant boom.

Furthermore, as already mentioned, there is the specter of dualism in any consideration of vital energy. But as I have shown in the preceding chapters, the problem of dualism is solved easily with a modicum of quantum thinking. It is time now for biologists and, with them, allopathic practitioners of conventional medicine to come to terms with reality—that the physical body is just a representation-maker for our subtle bodies and medicine must be extended to deal with all our bodies, not just the physical.

In Brief

Take the following ideas with you and give them further thought:

- One can make a good case for allopathic medicine only when it comes to medical emergency situations. The use of allopathy in other situations is suspect and choices must be examined carefully.

- Even allopathic procedures that we routinely accept today may be causing us harm. Immunization through vaccines is a case in point.

- Even biology, supposedly the parent science of conventional medicine, needs a paradigm shift. In the new aborning paradigm of biology based on the primacy of consciousness, the separateness of living beings from their environment is clearly seen as illusory. So we are not necessarily victims of the environment in its disease-producing capacity, as the allopaths would have us assume. It is time to reconsider the victim mentality with which we look at disease at the allopath's prodding.

• For our healing journey, which is ultimately a journey toward wholeness, means and end cannot serve separate goals. The techniques of allopathic medicine increase our separateness from the whole; by treating us as machines, it tends to make us into conditioned, choiceless machines. In contrast, alternative and complementary medicine, by paying attention to subtler, potentially more creative aspects of ourselves, tends to bridge our separateness from the whole. In choosing your proper medicine, consider this difference between the two.

PART 2

Vital Body Medicine

8

The Vital Body

Several years ago, when I was at a yoga research conference in Bangalore, India, I had the good occasion of watching a *pranic* healer in action. The Sanskrit word *prana* is best translated as "vital energy," a concept that Western medicine has discarded but one that is important not only in the Indian Ayurveda but also in traditional Chinese medicine, where it is called *chi*. The *pranic* healer makes the gesture of sweeping her subject's body in order to restore the balance of vital energy movements in the patient and thereby healing the physical body. It works more often than not. What gives?

Don't think that *prana* is an Eastern concept unknown to the Western mind. The eighteenth-century romantic poet William Blake wrote:

Man has no Body distinct from the Soul!
for that called Body is a portion of Soul
discerned by the five Senses,

the chief inlets of Soul in this age.
Energy is the only life and is from the Body;
and reason is the bound or outward
circumference of energy.
Energy is eternal delight.

The "energy" that Blake discovered to be eternal delight is not the energy that physicists talk about (the concept of energy did not enter physics until the 1830s), but is the *prana* of the Indian tradition and the *chi* of the Chinese.

The Vital Body and Psychophysical Parallelism

I have already introduced the vital body in earlier chapters (see chapters 3 and 4); this is a recap with additional insights about it.

The concept of vital energy was discarded in Western biology and medicine because of the implied dualism and because of the advent of molecular biology when it seemed that we could understand everything about the body through the chemistry of DNA and so on. But DNA alone cannot explain healing. As every physician and patient know, healing requires vitality, vital energy. Vital energy is not the product of body chemistry. Chemistry is local, but the feelings of vital energy, the feeling of being alive, is quite nonlocal. But then, where does vital energy come from?

A fundamental component of healing is regeneration. Even after a severe wound, the destruction of a massive number of cells, the body has the capacity of regenerating these cells exactly differentiated to perform the particular function as needed. If you say this happens due to cell division from neighboring cells, think again. The neighboring cells are often individualized differently.

As biologist Rupert Sheldrake has pointed out, regeneration is possible only because the blueprints of form come from nonlocal and nonphysical morphogenetic fields, and they provide the additional form-making needed for regeneration. The morphogenetic fields comprise the vital body. The vital energy that we feel is the movement of the vital body.

Molecules obey physical laws, but they know nothing about the contexts of living, such as maintenance and survival, let alone love or jealousy, that occupy us much of the time. The vital body belongs to a separate subtle world and contains the blueprints of form-making, forms that carry out the fundamental vital functions—the contexts of living. In other words, the vital body provides the body plans of the organs of the physical body that play out the vital functions in space-time.

The point is this: Physical objects obey causal laws, and that's all we need to know in order to analyze their behavior; I call their behavior law-like. Biological systems obey the laws of physics, but they also perform certain purposive functions: self-reproduction, survival, maintenance of integrity of self vis-à-vis the environment, self-expression, evolution, and self-knowledge.

Some of these functions you will recognize as instincts we share with animals. For example, fear is a feeling connected with our survival instinct, but can you imagine a bundle of molecules being afraid? Molecular behavior can be explained completely within physical laws, without giving it the attribute of fear. Fear is vital body movement that we feel, and concomitantly, a program guides the cells of a physical organ to carry out appropriate vital functions in response to a fear-producing stimulus.

The behavior of biological systems is interesting because these programs that run their functions are not related to the physical causal laws that govern the movement of their molecular substratum. I call this behavior program-like (Goswami 1994).

Rupert Sheldrake's great contribution to biology is to recognize the source of this program-like behavior. Sheldrake introduced nonlocal and nonphysical morphogenetic fields in biology to explain the programs that run biological morphogenesis—physical form-making for biological beings.

So again, the vital body is the reservoir of morphogenetic fields, the blueprints of form-making. The job of the physical body is to make representations of the vital body morphogenetic fields. The job of the representations is to perform functions of living, maintenance, and reproduction; the job of the vital body is to provide blueprints for representation-making.

It makes sense. If living forms are run by software programs, then the programs must have started somewhere from blueprints made by a programmer. Sure, the blueprints are now built into the hardware as form, and the program-like behavior of biological form is automatic. So it is easy to forget the origin of the program-like behavior and the programmer. But when the functions of built-in form go awry, what then?

So the vital body is needed. It has the original blueprints of biological functions, the morphogenetic fields that the physical body organs represent. Once the representations are made, the organs run the programs that carry out the biological functions. The representation-maker, the programmer, is consciousness. Consciousness uses the vital blueprints to make physical representations of its vital functions whose archetypes are codified in our supramental body, the body of laws and archetypes (see figure 11). When the quantum possibilities of a physical organ are collapsed by consciousness to the actuality of carrying out an intended biological function, consciousness also collapses the correlated movement of the corresponding vital body blueprint. It is this movement that we feel as a feeling.

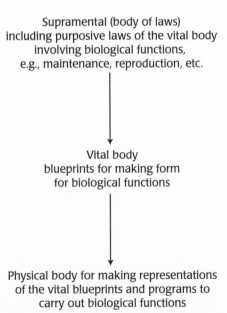

Supramental (body of laws)
including purposive laws of the vital body
involving biological functions,
e.g., maintenance, reproduction, etc.

Vital body
blueprints for making form
for biological functions

Physical body for making representations
of the vital blueprints and programs to
carry out biological functions

Fig. 11. How biological functions come down from Heaven (the supramental domain) to Earth (the material domain).

The objection of dualism is no longer tenable once you invoke quantum thinking. Consciousness can collapse, simultaneously and nonlocally, the possibility waves of *all the bodies* within consciousness—the physical, the vital, the mental, and the supramental. We have to assume that like the physical, the vital is also a quantum body.

What is *prana,* or *chi,* or vital energy? It is the quantum mode of movement of the vital body blueprint. When you are having the physical and mental experience of having an emotion, there is an extra, subtle, vital movement that consciousness collapses in your internal awareness; this is manifest *prana.* Is it possible to feel somebody else's feeling? You bet, through quantum nonlocality of the vital body (akin to mental telepathy); we call it empathy. Is it possible for a person of healthy *chi* to help another balance her *chi?* You bet, through quantum nonlocality.

Built into the Eastern philosophy of healing is that there are special people, spiritual healers, who can institute vital body healing by merely touching, or making a sweeping gesture with their hands over the body of the patient. Fortunately, this "hands-on healing" is not restricted to Eastern medicine; many spiritual traditions in the West use it, and people with special healing powers are revered in these traditions. Thanks to pioneers like Dolores Krieger and Dora Kunz, hands-on healing and Therapeutic Touch have made some inroads in modern Western culture.

What I was seeing at the conference in Bangalore in the episode cited at the beginning of the chapter, the psychic sweeping of the vital field of a patient, is quantum nonlocal transfer of healing vital energy, healing *prana.*

Emotions involve vital body movements in addition to the movements of the physical and the mental. Just watch yourself next time you are angry: The physical rush of blood to your face makes it red; angry thoughts arise—I will show them! But watch! There is something else, something more subtle, that you feel internally, that does not fall into either of these categories. That's the *prana,* the vital energy.

The vital body is indivisible; it has no micro-macro division, no structure. That is why feelings of the vital body are subtle, experienced internally. However, we acquire an individual vital body; it is functional, of course, through conditioning, as certain vital movements are conditioned to occur through repeated use, forming a pattern of individual habit.

Behold! Our physical and vital (and mental and supramental) bodies are separate substance bodies that go on in parallel, the parallelism maintained by consciousness. But don't picture the substances these bodies are made of as solid or concrete. This is not the quantum way of thinking of substances, not even physical substances. All substances are possibilities; only with collapse as actuality does consciousness give them all the substantiality they have. For the physical, the substantiality is often structural, quite concrete, like our individual physical bodies. For the vital (and the mental), even the individuality is functional, guaranteeing that it remains subtle all the way, even in manifestation.

The Evidence for the Quantum Nature of the Vital Body

Is there any evidence for the quantum nature of the vital body, which is the crucial hypothesis of the psychophysical parallelism that I am proposing?

Traditional Chinese medicine talks about pathways called meridians for the flow of *chi;* the acupuncture points used in Chinese medicine lie along these meridians (see chapter 10 for details). If the movement of *chi* is thus localized, at first considera-tion it may seem that the behavior of *chi* is classical and determin-istic. But Indians also have mapped the movement of *chi* (which they call *prana*) in pathways they call *nadis*. These *nadis* do not exactly coincide with the Chinese meridians. This is quite consis-tent with quantum behavior—the pathways of neither tradition are concrete, but mere guidelines for intuitive exploration.

So there may be an uncertainty principle operating between the localization and the direction of movement of *chi*. This is fur-ther confirmed by the fact that the Chinese characterize *chi* as hav- .
ing two complementary aspects of its wholeness (Tao), yin and yang, similar to the complementary characterization of material

objects as wave and particle. When we speak of "balancing the vital energy," in Chinese medicine, this means balancing the yin and yang aspects of vital energy.

The fact that vital energy, as in emotions, is felt in our private and internal awareness lends further support to an uncertainty principle operating in its movement. You cannot feel my feeling, because the collapse of my feeling in consciousness changes it (because of the uncertainty principle). (See also the discussion in connection with the internal nature of illness in chapter 4.)

The journalist Bill Moyers's TV series *Healing and the Mind* for PBS had a fascinating segment regarding Chinese medicine and the mystery of *chi*. In one segment, in answer to Moyers's question, "How does the doctor know he's hitting the right [acupuncture] point?" David Eisenberg, an American apprentice of Chinese medicine, said:

> It's an incredibly difficult thing to do. He asks her whether she feels the *chi,* and if she has a sensation, that's how he knows. He also has to feel it. My acupuncture teacher said it's like fishing. You must know the difference between a nibble and a bite.

And, of course, it takes years to learn to feel somebody else's *chi.* The feeling of *chi* is internal, normally not a part of our shared reality. How the acupuncturist shares in the *chi*-experience of a patient is like mental telepathy; it works through quantum nonlocality.

Very good evidence of the quantum nature of *chi* is coming from controlled experiments in China with *chi gong* (meaning manipulation of *chi;* also written Qigong) masters. These masters of *chi*-movement are asked to project "good" *chi* on plants, whose metabolic rate of growth is then measured and found to be enhanced. In contrast, when these *chi gong* masters throw "bad" *chi* on plants, the plants' metabolic growth rates are found to be adversely affected (Sancier 1991).

Clearly, the experiment shows a nonlocal connection between the vital bodies of the *chi gong* masters with those of the plants. Since nonlocality can never be simulated by classical machines (Feynman 1981), this is prime evidence of the quantum nature of the vital body.

Vital energy may become correlated with particular physical bodies or even places, however. In that case, the physical body or even the place may be said to "carry" the vital energy. This is most evident in such phenomena as the "phantom limb," in which one feels the limb even after it has been amputated. Perhaps this is also how radionics machines transmit vital remedies.

Can We Measure Vital Energy with Physical Instruments?

Can we measure vital energy with a physical instrument? The answer is no, by definition. Vital energy and physical instruments belong to two different worlds that do not directly interact. But there is a caveat.

The physical forms represent the vital body morphogenetic fields and are correlated with them. If we can measure these correlated physical forms as they change with the vital body movements, then indirectly we are measuring something about the vital body. This is what we do to measure thinking. Can we tell if somebody is thinking? Yes. We look at activity in the brain with a magnetic resonance imaging (MRI) device or with positron tomography.

I think maybe this is what the controversial technique of Kirlian photography does. Kirlian photography was discovered by the Russian scientists Semyon and Valentina Kirlian. It involves the use of an electric transformer called a tesla coil, which is connected to two metal plates. A person's finger is placed between the plates where a piece of film touches it. When electricity is turned on, what the film records is called a Kirlian photograph of the finger.

Typically Kirlian photographs show an "aura" around the object. Proponents of Kirlian photography claim that the color and intensity of the aura are descriptors of the emotional state of the person (whose finger is being used in taking the photograph). For example, a red and blotchy aura corresponds to the emotion of anxiety, a glow in the aura indicates relaxation, and so forth.

It is clear that some sort of energetic phenomenon is taking place. It has been verified that the energy involved cannot be controlled by the five senses. So originally, some researchers thought that what we were seeing were pictures of subtle energy flow from the finger to the film via psychokinesis. But this could only be if subtle energy were physical somehow.

An alternative materialist explanation has also been given—that the auras are related to sweating. Indeed, the presence of moisture in between the plates affects the photographs, creating controversy in their interpretation.

The reason I bring all this up is that it is possible to give a third explanation. Changes in vital energy as in mood swing do change the programs that run the organ representations whose functions also change, reflecting the mood swing. The photograph is measuring the change in the physical level, but because the physical level changes are correlated with the vital level changes, indirectly we are measuring the latter.

What Is Your Body Type?

If you ask an allopathic physician the question, "What's my body type?" he will say, "Well, that depends on your genes, doesn't it?" But does it? If you point out that genes are more or less just the instructions to make proteins, not for morphogenesis, the allopathic doctor will say in exasperation, "If genes cannot explain it, then there is no such thing as body type. And even if there were, it would not be important. Your disease does not care about how your personal body is constituted unless you have a genetic defect. The treatment does not depend on your so-called body type either."

But in medical systems that include the vital body—the morphogenetic fields that make form—body types make perfect sense. In this way, both Indian Ayurvedic medicine (see chapter 9) and traditional Chinese medicine (see chapter 10) have very important things to say about body types—the classification of our natural constitutions. They both give guidance about how to take care of our body depending on our body types, how for each body type diseases develop, and how the treatment depends on the body type. Vital body medicine is very individualized, and here is one of its greatest strengths.

Unfortunately, when you take a look at the two schools, the Indian and Chinese, at first you might be disappointed. The two systems don't always agree with the assessments of each other. Shouldn't science be monolithic?

So it is very important to note that since the vital body is subtle, vital body medicine also has to be subtle. We can, generally speaking, only have internal subjective experiences of the subtle. Thus we cannot, in general, expect to have a strongly objective, single observer-independent science for the vital body. At the same time, science requires at least weak objectivity—observer invariance; the conclusions drawn must be independent of a particular observer. We will see that the Chinese and the Indian systems do have enough points of similarity to satisfy the criterion of weak objectivity of vital body medicine. Once the cultural conditioning is taken into consideration, the two systems can be seen to complement, not contradict, each other.

The state of medicine for the vital body should make cultural anthropologists very happy. For some time, many cultural anthropologists have been challenging the idea of one set of universal laws for things, an approach that succeeds for the material universe. But now there is some consolation; at least in regard to the vital, and the interface of the vital and the physical, the cultural anthropologist may be right.

9

Ayurveda and the Healing of Vital Energy Imbalances

Ayurveda is the science of health and healing developed in India, where it has been in use for millennia. And thanks to its widespread use today in both India and abroad, and also thanks to brilliant expositors such as Deepak Chopra (2000), Vasant Lad (1984), and David Frawley (1989), Ayurvedic concepts such as the *doshas* have become more familiar in the United States.

Just to give you an example, I was at a party recently and this complete stranger asked me out of the blue, "So what's your body type? Are you a *vata,* a *pitta,* or a *kapha?*" Now *vata, pitta,* and *kapha* are names of Ayurvedic *doshas.* The implication was clear to me. Before this person would start a conversation with a stranger, she needed to know the Ayurvedic typology of the person, which, according to Ayurveda, is determined by a person's dominant *dosha.* Even a decade ago it used to be astrology—"Are you a

Sagittarius?"—that was used to break the ice between strangers. Astrology, move over.

But what is a *dosha*? Modern Ayurvedic physicians can tell you all about the imbalances of your *doshas* from your symptoms of unwellness but would be quite vague in defining them. They may say that physically these *doshas* are related to body humors, *vata* with intestinal gas, *pitta* with bile, and *kapha* with phlegm. If you are knowledgeable as to how medicine was practiced in earlier days in the West, you will recognize the importance of humors.

In the West, four humors were considered to be important. Choleric humor represented by yellow bile clearly corresponds to Ayurvedic *pitta*. The phlegmatic humor was represented by phlegm, corresponding to *kapha* in the Ayurvedic system. The other two were the melancholic humor as represented by black bile and the sanguine humor as represented by blood. These last two correspond to the humor of *vata* in the Ayurvedic tradition. Even today, our language says melancholy denotes a state of depression, which concurs with the Ayurvedic view that excess *vata* is most responsible for chronic diseases that may cause depression.

The medicine of earlier days also connected these *doshas* to the "five elements," a description of the material nature—earth, water, air, fire, and ether—then prevalent. *Vata* is air but needs the vessel of ether (empty space) to move; so *vata* reflects both ether and air. *Pitta* is clearly (digestive) fire, and since it needs the vehicle of water, it is seen to reflect the elements of fire and water. And *kapha* is water whose container is earth (the element, not the planet). So *kapha* reflects both water and earth.

But if this is all there is to Ayurveda, and you are a Western medicine practitioner, you will not be impressed. The old system is arbitrary, simplistic, and, of course, based on an archaic worldview. Compared to this, the worldview on which modern (allopathic) medicine is based is sophisticated. If one points out the empirical success of the Ayurvedic practice today, all the Western physician can do is shrug. There seems to be no way of understanding the importance of *vata, pitta,* and *kapha* within the current scientific worldview.

Many proponents of Ayurveda have begun to dispense with

the overarching model, or at least to de-emphasize it. They may say that these *doshas* are connected with the processes of the body as follows:

Vata: ordinary movement (such as circulation of blood)

Pitta: transformative movement (such as digestion)

Kapha: structure, or that which holds the structure together (such as the lining of the lungs)

These physicians recognize that the important thing is not to get bogged down with worldview questions. The human system may be much too complex to develop a proper theory of it that is able to connect all the way to the fundamentals on which the worldview is based.

Instead, these pragmatic Ayurvedic physicians use the concept of the *doshas* to classify all humans into seven types:

1. Pure *vata,* in which *vata* dominates the other two *doshas* (such people have a thin build and variable or changeable demeanor, just to mention two important characteristics).

2. Pure *pitta,* in which *pitta* dominates the other two *doshas* (giving us people of medium build and enterprising sharp intellect).

3. Pure *kapha,* in which *kapha* dominates the other two *doshas* (robust build, slow).

4–6. The three mixed *doshas, pitta-vata, pitta-kapha,* and *vata-kapha* (in whom the characteristics are combined—some of one and some of the other *dosha*).

7. The rare *pitta-vata-kapha,* in which all *doshas* are equally present.

In principle, there is also an eighth body type—the perfectly balanced one—but this is very rare.

The fundamental assumption of Ayurveda is that one is born with a given body type—a particular "base level" imbalance (called *prakriti* in Sanskrit). The contingencies of life, lifestyle, and environment take one to further imbalances *(bikriti),* causing disease.

Ayurvedic medicinal herbs, diets, and practices try to bring one back to the base level imbalance. Theoretically, it may seem desirable to try to correct even the base level imbalance, but that is very difficult to do and is usually not attempted.

In our formative years, when the structure is being built, *kapha* is said to dominate. In the middle portion of life, *pitta* dominates. In the declining years, *vata* tends to dominate. Since disease happens more when we are middle-aged and older, clearly most disease must be *vata* imbalance. Next most prevalent is *pitta* imbalance. *Kapha* imbalance is the least common. Our modern lifestyle also aggravates *vata*. So on the face of it, Ayurveda gives us a simple message: Watch out for that aggravated *vata*.

Nevertheless, the practice of Ayurveda is subtle. Even modern allopathic medicine has begun to emphasize lifestyle (the social theory of disease). For example, heart disease is associated with the so-called type A personality and lifestyle (hyperactive, overanxious, do-do-do) even by allopathic physicians. But there is a difference in the approach of Ayurveda, which takes account of the fact that not everybody with a type A personality gets heart disease. Ayurveda attempts to bring the patient back to one's base level *dosha* distribution, to one's *prakriti*. If the base level is already type A, Ayurveda does not bother to correct it. It is this individualized nature of Ayurveda that makes it so useful in treating chronic disease.

Nevertheless, major questions remain. Why does each one of us have a particular *prakriti*, a base level of *dosha* imbalance? Clearly Ayurveda, judging from its success, complements allopathic medicine, but how? Is there a scientific theory behind this way of looking at our body types that give us our tendencies toward disease, and healing?

Ayurveda and the Vital Body

At a deeper level, Ayurveda is said to be based on a more extended picture of ourselves than today's allopathic medicine (and modern Ayurveda by default), which recognizes us to be only the physical body. In addition to the physical body, Ayurveda allows us to be the movements of a vital body and, less important in the present context, the movements of a mental and supramental body as well, all grounded in consciousness, which is the ground of all being.

The physical body forms, the cells and organs of the body, are representations of vital body blueprints, which Ayurveda recognizes, and in so recognizing, Ayurveda is able to point out more ways that a person may get sick. Recall that in modern biology, Rupert Sheldrake has made the same point: Nonphysical morphogenetic fields (a more scientific name for the vital body) provide the blueprint for making physical form.

For allopaths, sickness involves chemical (and physical) functions of the physical body that have gone awry, and treatment likewise consists only of correcting the faulty chemistry (or physics). For modern Ayurvedic physicians, likewise, illness is at the physical level only, and imbalance considered for treatment pertains to physical features such as *vata, pitta,* and *kapha;* it's a bit more subtle than the allopathic approach.

But in deep-level Ayurveda, illness can also be the result of faulty nature, of faulty movements of the blueprints, of the vital body. Recall that these movements are associated with the programs that run the vital functions of the physical organs (the representations of the blueprints). Clearly, unless you correct the fault at the vital level, you can never make organ representations behave properly to carry out your vital function in a way that corresponds to physical health. Thus traditional Ayurveda at a deep level puts more emphasis on achieving a balance of the vital body qualities (the same is true of Chinese medicine), which in the Ayurvedic system are the precursors of the *doshas.*

This deep-level Ayurvedic scenario of illness has a clearer complementarity with allopathy (because it introduces the complementary disease-causing scenario involving the vital body) than does modern Ayurveda, which deals with the physical *doshas* alone. But there is a problem: What is the vital body in relation to the physical?

Here the traditionalist claims that the vital body is nonphysical. Unfortunately for the modern Ayurvedic physician, such a claim is precarious. Hasn't molecular biology eliminated all forms of vitalism or nonphysical life force? The postulate of a vital body also raises the specter of dualism and the question, How does a nonphysical vital body interact with the physical?

But this dualistic view of the vital body (and of consciousness itself) is a product of the myopic "single" vision of Newtonian thinking. In physics, we have replaced Newtonian classical physics with quantum physics, and the implication of this paradigm shift in physics is gradually being worked out for other fields of human endeavor such as medicine. In quantum thinking, vitalism is no longer a dualism. In other words, in quantum thinking we can postulate a nonphysical vital body without falling into the pitfall of interaction dualism.

As mentioned earlier, in quantum physics, all objects are waves of the possibilities of consciousness, which is the ground of all being; these possibilities can be classified as physical and vital (and mental and supramental). These possibility objects become the "things" of our experience when consciousness *chooses* out of its possibilities the actual experience in a particular quantum measurement (an event that physicists call collapse of the possibility wave).

Undeniably, an experience has a physical component, but if you look carefully, it also comes with a feeling that is its vital component (and thoughts and intuition are the mental and supramental components). This eliminates dualism, because the physical and vital can be seen to be functioning in parallel while consciousness maintains the parallelism (see figure 4).

So vital energies are what we feel in our vital body when we experience our physical organs. When these feelings are not "right," we feel illness. But we have a certain predisposition in our processing of vital energies. It is these vital predispositions that are the precursors of the physical "defects" called the *doshas* in Sanskrit.

Why Three Types of *Doshas?*

Consciousness collapses vital feelings along with the correlated physical organ (for the moment, let's leave out the mental and supramental aspects of our experience) for its experiences of living. We have to remember that our physical bodies are in constant flux; our cells and organs are being constantly renewed with the help of the food molecules that we eat. We also have to remember that the quantum possibilities of the vital body consist of a possibility spectrum (with a corresponding probability distribution determined by the quantum dynamics of the situation) of the morphogenetic fields, the blueprints of the vital body. The dynamic laws of the vital body functions and the contexts of its movement are contained in the supramental.

When we are first making representations, as when the one-celled embryo becomes a multicellular form through cell division, our choice of the vital body blueprint is free and the physical representation has a chance to correspond to optimal health under any internal and external environmental condition. In other words, though the vital body function and its laws, the supramental archetype, are always the same, we are free to choose the particular morphogenetic field, the blueprint that corresponds to that function.

This freedom of choice as to which vital blueprint will be used for physical representation-making is exerted according to the context of the external and internal environment of the particular physical body under construction. Of course, even at this stage, disease may arise because of (1) defects of the representation-making apparatus (the inherited genes), and (2) the inadequacy of the building material of the representation (our food intake and nutritional status). But given proper genetic endowment and proper nutrition, when we first build our physical body from the vital body instructions, we can be creative. And with creativity, we can even overcome some of the shortcomings of the genetic predisposition and/or malnutrition, and even environmental problems such as bacteria and viruses.

Thus it is that most children enjoy good health. However, some children suffer from bad health and this even without environmental contributions such as bacteria and viruses. Moreover, it is well known that diseases such as heart disease can be traced to indications appearing at an earlier age. What gives?

On our way to adulthood, by which time our physical bodies (the cells and organs) have been renewed many times, something called conditioning takes place, and this compromises our creativity. Conditioning is the result of the modification of probabilities to gain more weight for a previously collapsed possibility, a past response; it is due to the reflection in the mirror of memory. (See Goswami 2000, for more details.) With conditioning, we no longer possess the leeway to choose our vital blueprint to suit the environmental situation; instead we become predisposed to use the same blueprint that we have used before. If that blueprint is faulty, faulty representations will result.

There is one other alternative. Often, as we renew our body, we use more than one vital blueprint of a vital function to make the physical representation for that vital function; if so, both blueprints will be part of our vital repertoire. Suppose that in a subsequent case of organ-making, certain new environmental challenges are present that require a creative response. In that case, even though consciousness may not be able to choose a response (a vital blueprint) entirely outside the learned repertoire, it still may meet the environmental challenges partway by choosing a blueprint that is a combination of relevant blueprints learned in the past.

This kind of secondary creativity is sometimes called situational creativity (see Goswami 1999) in contrast to an entirely new response in a new context, which is called fundamental creativity.

In this way, there are three qualities of the vital body:

1. The ability to creatively shift the choice of vital blueprint for making a particular form depending on the context; this ability or quality of fundamental creativity is called *tejas* in Ayurveda.

2. The ability to choose a new combination of previously learned contexts of vital blueprints to make form; this is called *prana* or *vayu*.

3. The ability to respond via the most conditioned vital blueprint; this is called *ojas*.

All three are needed for the proper optimal functioning of the vital-physical bodies. Any imbalance will cause defects—the *doshas*, at the physical level. Corresponding to the three vital qualities, there are three *doshas* or defects of the representations. If there is too much *tejas* or fundamental creativity in the building of the body organs with the help of the vital morphogenetic fields, it results in the *pitta* type of body. Use of too much *vayu* or situational creativity at the vital body level gives us the *vata* type of body; and too much *ojas* produces the *kapha* type. The *doshas* are the "waste products" of how we use our vital body qualities to make form.

Another way of looking at this is to see *tejas* as the transformative vital energy an excess use of which produces a preponderance of metabolism in the physical makeup of the body and thus the *dosha* of *pitta*. Similarly, an excess of *vayu*—excessive and fickle vital movement within known contexts—translates as the *dosha* of *vata*, whose main characteristics are movement, fickleness, and changeability. An excess use of *ojas*—stable conditioned movements of the vital body—translates into the preponderance of the *dosha* of *kapha*, whose main characteristics are stability and structure.

There is a similarity in the three vital qualities, the *tejas-vayu-ojas* trio, to the three qualities of mind called the *gunas*. First, let's recognize that the Sanskrit word *guna* means "quality" in English. Second, later discussions will show that the mental *gunas* have the same origin as the vital *gunas*. They correspond to the three ways that the quantum mind can be used to process things: fundamental creativity, situational creativity, and conditioning (see chapter 14). Finally, note that the *gunas* (qualities) improperly used in out-of-balance fashion give rise to *doshas* (defects) at the physical level.

Where Does *Prakriti* Come From?

One question I will briefly deal with is about the origin of *prakriti* and why we have a self-styled *dosha* imbalance early on. Since *doshas* are the by-products of the more subtle vital qualities, we can ask: Are we born with an imbalance of these vital qualities? If so, why? The fact that many children suffer from chronic illness supports the view that there may be innate imbalances of the vital qualities of *tejas, vayu,* and *ojas* that we are born with. Why these imbalances? The succinct answer is: reincarnation.

I have previously introduced reincarnation (see chapters 3 and 6). Reincarnation is part and parcel of all Eastern systems of thought, and Ayurveda is no exception. The study of reincarnation falls within the purview of science since the demonstration of the existence of reincarnation proves the existence of the so-called subtle bodies— especially the vital and the mental (see chapter 3). The stimuli we experience in our lives and our responses to them produce brain memory. When a stimulus is repeated, the quantum probability of the response is more heavily weighted toward the previous response.

I call this quantum memory of the brain's response patterns. And since the mind is correlated with the brain, the mind also develops quantum memory of its conditioned habit patterns. It is this quantum memory, the modification or conditioning of the mind as we live it, that is reincarnated (see chapter 6). Now the inheritance of these modifications of the mind from previous lives, called *karma* in Eastern thought, has been empirically demonstrated (see Goswami 2001), giving further credence to the idea of reincarnation.

Traditionally, karma is understood as mental karma, mental propensities that we bring with us from our past lives. But a little thought shows that there can also be vital karma, vital propensities that we develop during our lives and that can then transmigrate to another life in the same way that mental karma is transmigrated (see Goswami 2001). The vital body is correlated to physical body organs at the chakras. Experiences and our response to them produce quantum memories for the organs that propagate to the vital body, giving rise to individual vital body propensities.

It is these vital propensities that we inherit from our past lives that give us innate imbalances of the vital qualities (fundamental creativity or *tejas*, situational creativity or *vayu*, and conditioned behavior or *ojas*) at the vital level. These innate vital propensities, imbalances notwithstanding, are as important in shaping our physical body as the conventional nature (our genetic endowment) and nurture (contribution of the physical environment).

Eventually, the innate imbalance of vital qualities that we are born with gives rise to *prakriti*, the natural base level imbalance of the *doshas*. Preponderance of *tejas* leads to the imbalance of *pitta*, and so forth, as described earlier.

The particular combination of *doshas* that we develop as we grow up, our physical *prakriti* or body type, is a homeostasis. Therefore, our physical body functions optimally when we remain at this homeostasis. If, however, our imbalances of vital qualities are not corrected and continue unabated, deviations from this homeostasis take place and disease is this movement away from the natural *dosha* homeostasis.

In this way, Ayurvedic healing can take two tracks. First, the obvious one: Correct the physical problems arising from the *dosha* imbalance beyond *dosha prakriti* at the physical level itself. Some of the Ayurvedic treatments are designed with this in mind—panchakarma, a cleansing of the body, for example. But this is only a temporary remedy.

The other track is to correct the imbalances of the vital body qualities. This correction alone can lead to a permanent remedy. This latter track can be practiced in two ways—passive and active. The passive path employs herbal medicine, administering herbs of specific patterns of *prana* to compensate what is missing. The active path is to transform directly the movements of *prana* at the vital body level. Breathing practices called *pranayama,* in which one watches the movements of breath along with the associated vital *prana* movements, are prime examples of the active path.

Doshas and Organ Sites

General considerations of the vital body as I have described give us an understanding of how *doshas* arise at the physical level. But how do we connect the *doshas* with the working of the body organs? We need further insights as to the nature of the vital blueprints or morphogenetic fields.

The ancient seers who discovered Ayurveda intuited something fundamental at this juncture (the founders of Chinese medicine had the same basic idea). It is now recognized that form in the macrophysical world exists in five different states: solid, liquid, gas, plasma, and void or vacuum. This has been obvious since antiquity, except the ancients had different names for the basic forms, such as earth, water, air, fire, and ether, generally referred to as the five elements.

The seers of Ayurveda recognized (paraphrased in the language of the quantum) that the vital body possibilities manifested in the same five basic states: earth, water, fire, air, and ether. Moreover, in the form-making of the physical body, the vital earth is correlated with the physical earth, vital fire with the physical fire, and so forth.

Now the *tejas* quality of the vital body is transformative, fundamental creativity. Thus it uses the vital fire, which correlates with fire at the physical level in the digestive system. Similarly, *vayu* is the empire-building situational creativity at the vital level and uses the air and ether aspects of the vital to represent movement in the physical, be it in the intestines or the blood vessels or the nerves. Finally, *ojas* is stability and uses the vital elements corresponding to solid earth and liquid water to make the physical elements with structure and stability.

If there is too much *tejas,* too much vital fire in physical form-making, there would also be the by-product of too much physical fire—a preponderance of the *dosha* of *pitta.* Similarly, too much *vayu,* too much use of vital air and ether results in the *dosha* of *vata;* and too much *ojas* means too much use of earth and water, which leads to the *dosha* of *kapha.*

So *pitta* resides mainly in the digestive system (stomach and small intestine) as excess acidity. *Vata* imbalance inhibits *pitta* and thus exists mainly in the large intestine (as intestinal gas), but also in the lungs and the respiratory system, the circulatory system, and the nerves. *Kapha* inhibits movement of *vata* and thus resides as phlegm mainly in the respiratory system and also in the stomach. In summary, the lower third of the body is thus said to be the domain of *vata,* the middle third the domain of *pitta,* and the upper third, the domain of *kapha.*

Normal Characteristics of People of the Three *Doshas*

In the foregoing I singled out a few of the most telling characteristics for each *dosha* to distinguish among them. But if you want to know which *dosha* or body type you belong to, a more detailed list is helpful. This is given in figure 12. If you are a mixed *dosha,* you will have a mixture of characteristics. Can you tell what your body type is from the list?

Vata	Pitta	Kapha
• Tall, rangy	• Medium, balanced build	• Heavy
• Lean, dry hair and skin	• Fine, straight (reddish)	• Smooth, thick skin
• Small eyes, irreg. teeth	hair (or bald)	• Thick, lustrous (coarse)
• Variable appetite	• Fiery eyes	hair
• Poor stamina	• Brisk appetites	• Large eyes, mouth,
• Light sleeper	• Good endurance unless	(even) teeth
• Fearful, axious	overheated	• Steady, stable appetites
• Spacey	• Angry, forceful, incisive	• Averse to activity
• Changeable	(cutting), impatient	• Cool, calm
		• Complacent

Fig. 12. The personality traits of people of the three *doshas.*

In my opinion, although such tests can tell you your dominant *dosha,* they cannot tell you your precise *prakriti,* the precise combination of *dosha* imbalances that is your particular body-homeostasis. You have to go to a good Ayurvedic diagnostician for that.

Dosha Imbalances

Why is knowing your *dosha prakriti* important? You are having your first lesson in individuality in the medical sense. For an allopath, you are not an individual, you are a machine for which only average or typical behavior can be given. In vital body medicine, you are an individual, a specific mixture of body structures and propensities called *doshas.* Not only that, you operate best when these *doshas* are close to their homeostatic base level values which are unique to you.

You can use your *dosha* information for taking care of yourself; all you need to know is what causes *dosha* imbalances from the homeostatic level of *prakriti,* and how to prevent these imbalances. This is the subject of the next few sections.

If you maintain your body according to your physical *prakriti* or base level imbalances, you breeze through life in good health. Problems arise when there is an imbalance, a disturbance in any of the *doshas* from this base level. Generally, the imbalance is most likely to be in your own body type, that is, a *vata* person is more likely to suffer from *vata* imbalance (overactive *vayu* at the vital level).

However, there is no strict rule; you can be a *kapha* person and still suffer from *pitta* imbalance. How can you tell if there is an imbalance beyond the original level of imbalance indicated in your *prakriti?* What causes these excess imbalances? One cause is seasonal change; the major one is lifestyle.

The *Dosha* Connection with the Seasons

There is a seasonal connection for the deviation of *dosha* imbalances from *prakriti* that can be predicated on the basis of the theory developed here. When it is hot in the external environment, as in the summer, that is the time for regeneration. Thus *tejas* is used plentifully, producing excess *pitta*. When it is cold, it means hibernation, and the stability of *ojas* is needed (excess of which produces an imbalance of *kapha*). If cold comes with dryness, the condition is right for *vata* imbalance. If it is cold and wet (rain or snow), all movement ceases, *ojas* dominates the vital body, and excess *kapha* is the result.

Throughout the year, if you are observant, you can see how seasonal changes affect your *dosha* imbalances. On the East Coast of the United States, when the winter is cold and dry, it is the excess *vata* that makes people enjoy movement even though it is cold. But in early spring, when the weather turns cold and wet, it is time for excess *kapha* imbalances, and one tends to catch cold (which most often is due to excess unbalanced *kapha*). When it is hot and humid in the summer, you can easily notice, especially if you are a *pitta* type, that excess *pitta* is causing you problems with acidity. So in summer, we all prefer cool food and drink, but for *pitta* people, these are a must.

In general, if your body type matches your environment, you have to be extra vigilant about keeping the imbalances from going out of control.

Vata Imbalance and Its Remedy

If you are a *vata* person balanced in your *prakriti,* you are cheerful, enthusiastic, and full of do-do-do energy. And why not? Whatever changes take place in your life situation, the reservoir of learned contexts of situational creativity at the vital level is able to reestablish your physical body homeostasis.

If, on the other hand, your active life is full of anxieties and worries, your body is full of aches and pains, and even your quality of sleep has given way to restlessness, then ask: Is my *vata* still in balance? Is my *vata* still near the base level of my *prakriti?* Those symptoms are symptoms of *vata* imbalance irrespective of whether you have *vata* dominance or not.

One scenario for *vata* imbalance is common to everyone: As we age, *vata* tends to be aggravated. It is just part of aging. With this kind of *vata* aggravation, there is not much we can do. There is sleeplessness, some loss of memory (here is everybody's chance to become the "absent-minded professor"), some aches and pains. Even the appetite will not be like it used to be.

But there are other scenarios. Suppose the change in your life situation is drastic, so drastic that the reservoir of *vayu,* those learned contexts of situational creativity, is not adequate to make the vital level adjustments in a hurry. So *vayu* is going to be over-worked, causing a lot of *vata* fallout at the physical level. Situations like this arise in our life when we travel, when we move our residence from one city to another, when we change jobs, at times of divorce or the death of a spouse.

I know. A few years back, in the course of a year, I got divorced, I started courting another woman (whom I later married), I changed jobs, and moved from one city to another much bigger city. On top of this, I got a grant so there was pressure for mental accomplishment that I had not encountered in years. And I was already struggling with aggravated *vata* due to advancing age. Can you imagine the degree of *vata* aggravation all this caused? I was becoming so disoriented that the following year I had three auto accidents in a period of six months.

So what's the remedy? Ayurvedic medicine suggests several tracks: proper diet, herbal remedies for *vata* aggravation, an environment of warmth and moisture, oil massage, certain hatha yoga exercises, a cleansing process called *panchakarma,* and relaxation. The details of diet and herbal remedies can be found in any good book on Ayurveda.

One advantage of Ayurvedic medicine is that it is mostly common sense. Unless you neglect your imbalance so long that the aggravation becomes severe (in which case the imbalance will lead to the physical symptoms of what we normally call a disease), it is quite possible to use the tracks in the previous paragraph as preventive medicine. You need never go to a doctor.

In my case, my usual regimen of yoga and meditation was unable to cope with the degree of *vata* aggravation. I never did solve my *vata* imbalance while living in the big city. Fortunately, the movement of consciousness cooperated, and it was necessary to move away from that city. Within six months, my *vata* became balanced. The main thing was the relaxed lifestyle of my new habitat. I should mention, however, that my wife helped in several ways: diet with fresh vegetarian food full of *prana*; long walks in nature; a lot less thinking and much more laughter. One problem concomitant with too much *vata* is too much mental work and the resulting tendency of taking yourself too seriously; you become full of (hot?) air.

Pitta Imbalance and Its Remedy

If you are a *pitta* person, you exude creativity and you are intense. When *pitta* is balanced you are able to handle your natural intensity with joy because you obviously like being intense. But if the intensity is there and the joy is missing, then *pitta* is unbalanced.

Do you see how it works? *Pitta* is a side effect of overworking *tejas,* of fundamental creativity at the vital level. *Tejas* helps us build a good digestive system and maintain it with proper renewal as needed. But if the drive, the intensity, becomes too much, *tejas* is overworked and the result is *pitta* imbalance.

A common scenario is in our middle years. We have stopped growing, so the pressure on the digestive system and *tejas* at the vital level is considerably reduced. Unfortunately, the inertia of habit keeps them going at the same level as at a younger age. This overworking of *tejas* continues until we settle down in the last third of our life. So in our thirties we have to accept a certain aggravation of *pitta* which results in acidity and heartburn, thinning of hair, vulnerability to stress, and things that take away the joy of intense life.

There are foolish ways to aggravate *pitta,* such as overworking the digestive system unnecessarily by eating improper food. When we are young, our digestive fire is strong and it is considerably intensified by eating hot and spicy foods. The *tejas* of the food is all used up for a good cause—growing a healthy body. But when you don't need such intense digestive fire any more, then overworking the system with unneeded *tejas* will produce *pitta* imbalance. If you still fail to take notice, you will end up with an ulcer.

Organization is not a forte of creative people of the *pitta* type. But when such organizational demands are made, the system reacts with anger, frustration, or resentment, the expression of which requires *tejas*. The excessive use of *tejas* gives the by-product of excess *pitta* at the physical level. So stress is a *pitta* aggravator. If you don't watch it, heart disease may result.

The remedy for *pitta* imbalance is moderation. Reduce the intake of stimulants such as coffee. Meditate. Take long nature walks. Let that extra intensity be used up in the appreciation of beauty.

Kapha Imbalance and Its Remedy

As a *kapha* person, your strong point is strength and stability, which endow you with generosity and affection to give to others, and this giving makes you happy. A *kapha* person is able to live a long happy life, yet a few glitches may arise.

In our childhood, the body is building itself; *ojas* is needed in abundance, and occasionally this results in excess *kapha*. This gives a child susceptibility to colds, sore throats, sinus trouble, and so forth; this susceptibility continues for the rest of the person's life in an otherwise healthy existence. One does not have to be a *kapha* person to contract this particular consequence of *kapha* imbalance.

But after our bodybuilding phase is over, vital strength and stability, *ojas,* with nothing to do has a tendency to produce obesity. This is the sign of *kapha* imbalance, which can lead to other imbalances if not controlled. Since our culture does not approve of obesity, insecurity results from it. If, in spite of this insecurity, one continues to be generous and giving, it will produce clinging. On the physical side, obesity puts too much stress on the heart and leads to hypertension and difficulty in breathing.

Still another scenario is faulty diet heavy in sweet foods. One manifestation of this route to *kapha* imbalance is diabetes.

Whereas treating *vata* imbalance requires a ho-hum life, dealing with *kapha* imbalance requires the opposite: more stimulation and variety to shake up the inertia. *Kapha* imbalance treatment also requires weight control, avoidance of sweets, and an exercise routine.

Panchakarma and Herbal Medicine for Which You Need a Doctor!

In Ayurveda, great emphasis is put on periodic cleansing of the systems of our body to get rid of excess humors due to the imbalances of *vata, pitta,* and *kapha,* also called *ama.* For example, *pitta* imbalance will create *ama* in the intestines. This can be dealt with through periodic cleansing of the affected organs. *Panchakarma* consists of five such cleansing procedures: therapeutic sweating, nasal cleansing with or without herbs, purgation of the stomach and intestines achieved through herbs or enemas, oil massage, and bloodletting. *Panchakarma* requires the supervision of a trained Ayurvedic physician.

As mentioned earlier, prescription of Ayurvedic herbs requires a trained physician. Reading books on Ayurveda helps more as preventive medicine and less for healing of an already aggravated condition.

The standard for Ayurvedic herbal medicine is very high. To quote from Charaka, one of the authorities:

> A medicine is one which enters the body, balances the Doshas, does not disturb the healthy tissue, does not adhere to them, and gets eliminated out through the urine, sweat, and feces. It cures the disease, gives longevity to the body cells, and has no side effects (quoted in Svoboda and Lade 1995).

A reminder. Ayurvedic healing must be individually designed. Ayurvedic medicine must be prescribed not only with your body type in mind, your *dosha prakriti,* but also your lifestyle, and your personality. So it is best not to use it as self-help, but to get the help of a trained physician.

I heard a story that illustrates this brilliantly. A brahmin (a Hindu scholastic person) of small physical stature was invited to a king's palace for dinner and he could not help himself and overate. So he came to the king's (Ayurvedic) physician asking help with digestion. The physician gave him a pill but warned him.

"Look, this is very potent medicine that I give to the king himself when he sometimes complains about overeating and digestion. One pill will be too much for you. Cut it in quarters and take only a quarter. Okay?"

"Okay," said the brahmin.

But when he came home, he had second thoughts. As was customary, he was given some leftovers from the dinner, which he had brought with him. "If I take the whole pill, surely I should be able to digest not only what I have in my stomach now, but also all the rest of the food." So he ate the rest of the delicious food, took his digestive pill, and went to sleep.

In the morning, the Ayurvedic physician was making the rounds with his son; it just so happened that they were in the neighborhood where the brahmin lived, and the kind physician thought of inquiring how the brahmin was doing. The door to the brahmin's house was unlocked. The physician and his son went in; no one was there. They went all the way to the bedroom and knocked; still no one responded. The physician pushed the door open, and lo! There was nobody on the bed, only some clothes. On examination, the clothes were found to be full of feces.

The physician took one look and realized what must have happened.

"Oh, my God!" he exclaimed.

"What's the matter, father?" The son did not understand.

"The poor brahmin must have taken the whole pill. Look, he digested the food, no doubt, but the pill digested him, also," said the father, pointing to the feces.

In Summary

Take the following ideas with you:

- Ayurveda is the Indian version of vital body medicine according to which disease is due to the imbalance of vital body movements called *prana,* vital energy.

- Ayurveda is medicine designed for the individual. According to this system, every individual is characterized by a combination of certain body "defects" called *doshas* in Sanskrit. There are three *doshas: vata, pitta,* and *kapha.* The dominant *dosha* determines an individual's body type, according to Ayurveda.

- Ayurveda is quantum medicine. This becomes clear as soon as you attempt to understand why there are three types of *doshas.* There are three types of *doshas* because there are three ways we can process the movements of a quantum body: (1) via fundamental creativity (in the case of the vital body this quality is called *tejas*); (2) via situational creativity (for the vital body, this quality is called *prana* or *vayu*); (3) via conditioned habit (for the vital body, this quality is called *ojas*). Each of the *doshas* is associated with one of the ways we process vital energy: *Pitta* is associated with fundamental creativity or *tejas, vata* with situational creativity, and *kapha* with conditioning.

- How do the *doshas* originate from the imbalance of vital energy movements? Excessive use of *tejas* or fundamental creativity for processing vital movements creates the physical body defect or *dosha* of *pitta.* Excessive use of situational creativity, *vayu,* gives rise to the *dosha* of *vata.* And excessive use of conditioning, *ojas,* creates the imbalance of *kapha.*

- For a quick orientation, think of *vata* as the tendency to produce excess intestinal gas, *pitta* as the tendency to produce excess acidity in the stomach, and *kapha* as excess phlegm in the respiratory system. For details, consult figure 12.

- With the application of figure 12, you may already have a valuable piece of information, your body type, which can be pure *vata, pitta, kapha,* or a mixture of two or more. Determine your body type.

- Everyone is born with a certain homeostatic base level of the three *doshas*; this base level homeostasis is called *prakriti. Prakriti* originates from vital body habit patterns that we accumulate in many incarnations, according to Ayurveda.

• According to Ayurveda, disease is excess or deficit imbalance of one or more of the *doshas* compared to their base level value of *prakriti.*

• Become familiar with what causes *dosha* imbalance away from *prakriti,* and consider the remedies. Apply this knowledge to your particular case.

• Once again, realize that your vital body functions best when your quantum creativity and classical conditioning are in balance in how you use your vital body. Now think how you can bring these two aspects of your nature into balance. See also chapters 10 and 17.

1 0

Vital Energy Imbalances and Their Healing in Traditional Chinese Medicine

Chinese medicine is based on ideas very similar to those of Ayurveda, but how the ideas are put to work is quite different. The Chinese noticed that the basic entity that we feel in the vital body, the vital energy that the Chinese call *chi,* has a complementarity to it. They called the two complementary aspects of *chi* "yang and yin," and you can see quantum insight in the Chinese view.

Yang is the transcendent, wavelike character of *chi:* expansive, nonlocal, creative, heavenly (analogous to *tejas* in Ayurveda). Yin is the immanent particle-like character of *chi:* contracted, localized, conditioned, and earthly. (Yin is analogous to both the concepts of *vayu* and *ojas* in Ayurveda.) Both yang and yin aspects are needed to express the full potency of *chi.*

The underlying philosophy of Chinese medicine is Taoism, with its emphasis on the twofold complementarity of the yin and yang, rather than the threefold characteristic of the *gunas* and *doshas* of Indian philosophy. So the Chinese characterization of body types is twofold: yang-type, those that represent the yang component of the vital energy; and yin-type, those that represent the yin-component.

The twofold distinction of yin-yang at the vital level gives rise to the well-known twofold opposites of the human body as representations of the vital. The yin-yang duality is effective for differentiating body types: cold-hot, moist-dry, heavy-light, slow-rapid, passive-aggressive, still-active, stable-creative, inward/downward-outward/upward, and so on.

You can see that the threefold differentiation is even more effective. For example, consider the cold (yin) and hot (yang) pair of opposites. In Ayurveda, hot would correspond to *pitta* type, but cold has two possibilities: *vata* and *kapha*. *Vata* is cold and dry (the air quality), and *kapha* is cold and moist (the water quality). Thus more information is conveyed with the threefold characterization.

But the Chinese make up the deficiency of the twofold yin-yang differentiation of the body types with an elaborate mapping of the organs of the body combining the yin-yang theory with a Chinese version of the five elements theory.

The Organs in the Chinese System

So far so good. But when the organs are introduced in the theory of Chinese medicine in conjunction with the five elements theory, everything becomes a bit vague and confusing. If you read any modern discussion of traditional Chinese medicine, five yin organs will be mentioned along with five yang organs. Moreover, you will be told that meridians, the channels for the flow of *chi*, connect these organs. No wonder that modern researchers are then tempted to look for these meridians (which play an essential role in acupuncture) in the physical body. But nobody has ever found any channel of flow in the physical body that can be called a meridian, nor has anybody ever found in the physical body any energy, subtle or gross, that is akin to *chi*. Both *chi* and meridians belong to the vital body; that is why all efforts to find them in the physical body end in failure.

It is imperative, then, that we develop a language that avoids this confusion. To describe correctly what is happening, our language concepts must refer to both the vital body (consisting of the morphogenetic fields or blueprints or body plans and the modes of their quantum movement, *chi*) and its physical body representations (the organs). The result will read a little more complex than the usual fare, but we will be able to avoid all misunderstanding.

We are now ready to introduce the clever part of traditional Chinese medicine. It defines the morphogenetic fields and their interrelationships through the five elements in the Chinese version—earth, water, fire, metal, and wood. This version is not entirely satisfactory, philosophically speaking, but it can be taken as a phenomenological coup. In other words, since we don't know how to classify the vital morphogenetic fields except to look at what physical organs they correspond to, we may as well choose the five elements according to the relationship that we empirically see between the physical organs (including their processes) of our body—or on the basis of what we see in physical nature.

The ancient Chinese saw nature exhibiting a circular relationship among its five elements. Two such relationships are recognized in Chinese philosophy: promotion (also called mother-son relationship) and control. For example, wood promotes fire. But with water, fire is extinguished—an example of control. The Chinese recognized the same relationships in our physical body; for example, the liver promotes the heart, but the heart controls the lung.

The ancient Chinese had the brilliant idea of classifying the morphogenetic fields of the vital body into five types according to the five elements: earth, water, fire, metal, and wood. Each type has complementary yin and yang movements. Therefore, each has two classes of organ representations. The class of organs associated with yin movements is called *zang* organs; the class associated with yang movements of its vital blueprint is called *fu* organs.

Each of the organ representations of a class simultaneously has promotion and control relationships with other members of the class. This means feedback, a circular relationship.

An example will clarify the circularity of the Chinese way of setting up relationships. Fire is controlled by water, but fire cannot directly control water. It can, however, promote earth, which *can* control water. Thus feedback is indirectly allowed, giving circularity.

What mediates the relationships that we are talking about, promotion or control? The flow of *chi,* of course, through the meridians connecting the vital blueprints.

Five *zang* (physical) organs represent the yin aspect of the five types of morphogenetic fields (wood, fire, earth, water, and metal). Similarly, corresponding to the yang aspects of the five types of morphogenetic fields in the vital body, there will be five physical *fu* organs, each corresponding to one of the elements.

Since yin represents earthly stability, the *zang* organs that manifest from the yin aspect of the vital blueprints are "full," but the fullness is dynamic, not stagnant. Generally, these organs store stuff, but the stuff can come in and go out. These organs (organs of storage) are: liver, heart, spleen, lung, and the kidneys.

Since yang represents the heavenly movement including creative movement, the *fu* organs, which are the representation of the yang aspect of the vital body plans, are the organs that either take in or eliminate. In other words, they are organs of transfer: gallbladder, small intestine, stomach, large intestine, and urinary bladder.

Notice that for each *zang* organ we can find a *fu* organ that is external to it. This is a *zang-fu* (internal/external) pair of organs that represents the yin-yang aspects of each of the five types of morphogenetic fields in the vital body:

• Liver and gallbladder correspond to vital wood.

• Heart and small intestine correspond to vital fire.

• Spleen and stomach correspond to vital earth.

• Lung and large intestine correspond to vital metal.

• The kidneys and urinary bladder correspond to vital water.

The promoting relationship among the *zang* organs can be illustrated as follows: As water nourishes wood, the kidneys (which represent the yin aspect of the vital water type morphogenetic field) nourish the liver (which represents the yin aspect of wood) by supplying it with essence.

As wood nourishes fire, the liver (representing the yin aspect of the morphogenetic field of the type wood) nourishes the heart (which represents the yin aspect of the morphogenetic field of type fire) by supplying the heart with blood.

As fire nourishes the earth, the heart (representing fire) nourishes the spleen (which represents earth) by providing it with heat from the circulation of blood.

As earth nourishes metal, the spleen (representing earth) promotes the lung (representing metal) by providing it with the essence of food.

As metal promotes water, the lung (representing metal) promotes the kidneys (representing water) by providing them with water (in descending movement).

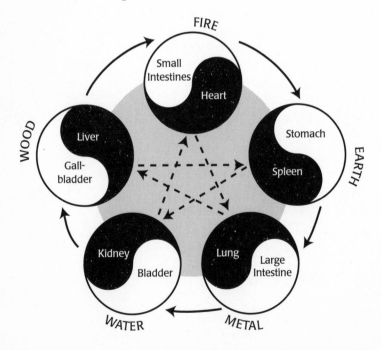

Fig. 13. The *zang-fu* organ pairs and their promoting (solid lines) and controlling (dotted lines) relationships.

Figure 13 shows the vital elements, their physical representations (the *zang-fu* pairs), and their promoting and controlling relationships. Using figure 13, you can make use of the controlling relationships among the vital elements to figure out the controlling relationships of the physical organs. For example, the quenching helping hand of the kidneys (water), since water controls fire, can control a flaring-up of the heart (fire). Likewise, as metal can cut wood, a flaring-up of the liver (wood) can be controlled by the lung (metal).

In this way, we find that both traditional Indian and Chinese systems use general principles about the vital morphogenetic fields to:

a) define health, which means balance and harmony

b) define disease, which means the absence of balance and harmony

c) derive relationships between the systems of the physical body, which then can be used to develop cures.

Disease and Healing

Disease in Chinese medicine means imbalances in yin or yang and also of the vital energy *chi* level of the morphogenetic blueprint of the ten *zang-fu* organs (see figure 13). The *chi* at each vital blueprint corresponding to an organ is denoted by the organ name followed by *chi;* for example, lung *chi* refers to the *chi* at the morphogenetic blueprint of lung, metal; however, the same *chi* is also called metal *chi*. There can be excess or deficiency of *chi* corresponding to an organ; there can also be stagnancy. All three conditions need to be corrected.

Traditional Chinese medicine holds that the vital body has the ability to resist pathogenic factors and repair the physical body. This ability is called antipathogenic *chi*. The history of a disease, then, is the history of the struggle between the pathogenic factors and the antipathogenic *chi*. The main factor in the struggle that antipathogenic *chi* engages in is to adjust the balance between its yin and yang, the conditioned and creative aspects of its movement. So disease can be talked about in terms of the imbalance of yin and yang.

Consider, for example, catching a cold after exposure to a cold wind. Exposure of the physical body to the cold wind produces a preponderance of yin at the vital blueprint corresponding to the skin and muscles adjacent to it. Since the skin is in external/internal relationship with respect to the lungs, an energy imbalance at the vital blueprint of the skin may also affect the vital blueprint of the lungs (which is metal) and cause an imbalance there as well. Excessive yin of the vital blueprint of the skin will produce deficiency of yang for the metal counterpart of the vital body. Thus the lungs (the physical representation of the vital metal) may not be able to perform their normal function of nourishing water. An excess of fluid and phlegm will result—the symptoms of the common cold.

A cold also can be the result of an internal pathogenic situation. Suppose there is a long illness that has depleted the yang at the vital level. Yang deficiency translates as underfunctioning of physical body organs, which reduces the resistance of the body and makes it susceptible to catching cold. Similarly, a preponderance of yang produces an overfunctioning of the affected organ, resulting in heat symptoms such as a fever.

Disease can also result from the preponderance or deficiency of yin and can be discussed from that point of view.

The Chinese medicine practitioner employs special herbal medicines and diet (and also massage and acupuncture, discussion to follow) to correct the energy imbalance at the affected organ. If the affected organ is lung, which is a metal organ, food and herbs connected with metal *chi* are given.

Some of the herbs used in traditional Chinese medicine have natural chemicals that resemble the synthetic chemicals found in allopathic drugs. So there is a tendency to regard these herbs (and also herbs used in Ayurveda) in terms of their chemical and physiological effects alone. But this misses the important vital energy aspect of herbal medicine. Whereas the herbs of the Chinese medicine act on *two levels,* both physical and vital, when the active chemicals of such herbs are isolated, only their physiological effect remains, and we have lost something.

Does Chinese herbal medicine work? Western doctors collaborated with Chinese doctors to design a trial using a group of children suffering from the skin condition of eczema. The doctors prepared a "sham" tea made of a set of traditional herbs having nothing to do with the treatment of eczema and compared the effect of this with that of "real" tea made of the correct herbs prescribed according to the principles of traditional Chinese medicine.

Half the group of children (randomly chosen) were given the real tea for an eight-week period, followed by a washout period of four weeks, and then another eight weeks of "sham" tea. The other half got the treatment in reverse order, sham tea first, then real tea. The results were dramatic. Whenever the children got the real tea, they saw a great improvement in their skin condition; whenever they got sham tea, their skin condition deteriorated dramatically (Sheehan and Atheron 1992).

Like Ayurveda, and unlike allopathic medicine, treatments are individualized in Chinese medicine. This is because two individuals may both suffer from the same disease, say stomach ulcer, but the imbalances that gave rise to the ulcer may be very different in the two cases. By the same token, if two people have the same imbalances in the movement of *chi*, they can be treated in the same way, irrespective of their symptoms.

Of course, the best treatment when a physical organ is affected due to excess or depleted *chi* at one of its vital blueprints is to direct balancing *chi* to the latter from another vital blueprint. Traditionally, this is the most spectacular aspect of Chinese medicine, and currently the most famous in countries other than China. Of course, I am talking about acupuncture.

Acupuncture is healing by puncturing the skin at various points with tiny needles. Why should such a simple procedure heal? Also, the place you apply the needle may have no spatial relationship to where the problem is. For example, an acupuncturist may apply needles to the big toe to cure a headache.

Acupuncture—How It Works

It is believed that acupuncture was discovered as a by-product of war. Warriors who were injured by the enemy's arrows made the discovery. They found that although an arrow lodged in the body hurt, it also relieved chronic pains that today we would associate with arthritis or tendonitis. The legend also says that when the reports of the soldiers reached Taoist sages who must have been experts in Chinese medicine, they realized what was happening. In the spirit of science, they pierced their own bodies with needles and mapped the pathways of *chi*, the meridians.

As mentioned before, the principal aspect of the theory of acupuncture is that there are channels for the flow of vital energy among the vital blueprints that are the reservoir of the programs that run the biological functions of the organs. These channels are called meridians. Meridians are channels for the flow of vital energy between the blueprints of organs, to put it simply.

For details, I will follow the exposition of Professor Yen-Chih Liu (1988) of Beijing College in China. There are 12 meridians correlated with the body and each is associated with either a *zang* or a *fu* organ, whereby arises its importance. Why 12 when there are only ten *zang-fu* organs? Chinese medicine recognizes two other vital blueprints with names, the Triple Burner and the Heart Protector, for which no form exists, but they have an important role in the transmission of *chi* between organs.

The second most important aspect of the theory is that there are places on the skin where the functioning of these principal channels can be influenced; even the channels of the Triple Burner and the Heart Protector have specific external points associated with them. These are places where outside influences can affect the organs and their blueprints, and the Triple Burner and the Heart Protector. The bad news is that external pathogens (for example, a gale, a cold wind) can affect an internal organ (and its blueprint) through these areas. But of course, this is also the good news. We can apply external energy for healing to an internal organ (and its blueprint) through the same points. These are the acupuncture points.

Massage therapy can also be applied to the areas surrounding acupuncture points. In fact, according to some Chinese experts, when acupuncture was first discovered, practitioners used only their fingers to influence the movement of *chi*—never mind the discovery story that tells otherwise. Today, the practice of manipulating *chi* with fingers is called acupressure.

Acupuncture became famous in the United States (it was already in vogue in Europe) in 1972. In the Nixon era, a journalist with the first American delegation to China underwent an appendectomy while there, without anesthetic but with the help of pain relief provided by acupuncture. Ever since, allopathic physicians have looked for a physical explanation of the meridians; for example, many allopathic physicians think that the meridians may be related to the nervous system.

To remind you once more, the meridians are not physical channels at all, nor is anything physical moving through them. Instead, they belong to the vital plane and give us the approximate pathways of movement of vital *chi* between the vital blueprints (the morphogenetic fields) of the organs of importance (the *zang-fu* organs). This should be obvious when we realize that some of the meridians connect two entities, the Triple Burner and the Heart Protector, which are conceived of as purely vital entities (morphogenetic fields with no physical representations).

Once the vital blueprints, the vital source of the programs that run the functions of the organ representation, are restored in terms of balance and harmony, the restoration of the organ functions follows quickly.

Why do the meridians describe only approximate pathways? Because ultimately, vital energy is quantum in nature and therefore it is impossible to describe its movement via exact trajectories. This is a dictum of Heisenberg's uncertainty principle.

How acupuncture is actually practiced today supports the quantum view of vital energy. Although traditionalists insist that the meridians are quite fixed, as are the acupuncture points, they agree that rather than being points, they denote areas. Today, some acupuncturists do not necessarily use the traditional meridians and acupuncture points. They ask the patient or even use methods such as muscle testing (a technique of applied kinesiology) to pinpoint where to apply the needles in order to heal the diseased organ.

I will tell you my experience with acupuncture. I was suffering from a stiff muscle pain in my left upper arm caused by a fall. It was a pain that wouldn't quit even a month after the accident. It happened that I was at the Sivananda Ashram in Val Morin near Montreal, Canada, giving a series of talks when I met Dr. Gopala, an American acupuncturist, who was also giving talks there. One thing led to another, and I happened to mention the pain in my arm, and Dr. Gopala asked if I would like an acupuncture treatment. I hadn't thought about it (this was before I became active in researching Integral Medicine), but I was curious and agreed to a treatment. The first treatment got the pain down considerably; and a second treatment after two days led to a complete cure.

When I had my acupuncture treatment, Dr. Gopala used muscle testing. He would poke places in my left arm and check my right arm muscles for strength or weakness; if the muscles showed strength, he would consider that a hit. That would be the place he would puncture with the needles.

How does acupuncture work for relieving pain? For a body with healthy *zang-fu* organs, the application of the needles to suitable areas can stimulate the general level of yang *chi* (manifest *chi*) to the body, especially to the brain areas that produce endorphins, the brain's own opiates. The manifestation of the vitality of *chi* at the vital level manifests brain states with endorphins. Indeed, narcotic antagonist drugs that block the action of opiates can neutralize the healing effect of an acupuncture treatment.

Acupuncture is designed to heal many other ailments, not only pain. As stated earlier, one can use acupuncture to allow vital energy to flow between the vital blueprints of any two *zang-fu* organs to correct energy imbalances to produce healing.

Figure 14 shows one of the principal channels, the Lung channel, with both the internal and the superficial parts of the pathway. Notice that the superficial branch of the meridian on the arm passes a point above the radial artery of the wrist. This explains how the Chinese medicine practitioner can diagnose disease by reading the pulses, which is a highly sophisticated art in traditional Chinese medicine.

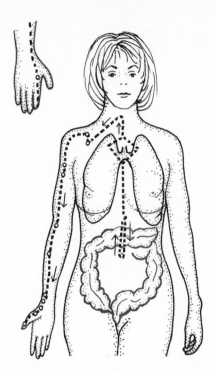

Fig. 14. A principal meridian, the Lung meridian.

Is There Any Possibility of Integrating Ayurveda and Traditional Chinese Medicine?

Some authors, notably Robert Svoboda and Arnie Lade (1995), have already begun the exploration of the tantalizing possibility of integrating the two great Eastern systems of medicine— Ayurveda and traditional Chinese. These authors regard yin, yang, and *chi* of Chinese medicine as a triad of body types similar to the three *doshas* of Ayurveda. In the quantum view (I think, in the view of any nondual philosophy), yin and yang are the dual, complementary aspects of *chi;* it cannot be otherwise. So in the Chinese system only a twofold body type can be formulated.

Actually, there is more similarity in how the two systems incorporate five elements theory in their schema. I mentioned before that the Chinese use of the five elements is very clever in that it is consonant with a circularity of support and control among the organs (see figure 13). What is not so well known is that there is such a thing as a wheel of support and wheel of control in Ayurveda also (see figure 15).

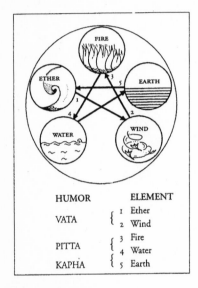
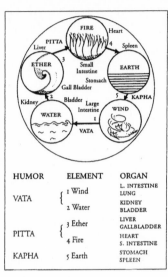

HUMOR	ELEMENT	
VATA	{	1 Ether
		2 Wind
PITTA	{	3 Fire
	{	4 Water
KAPHA	{	5 Earth

HUMOR	ELEMENT		ORGAN
VATA	{	1 Wind	L. INTESTINE LUNG
		2 Water	KIDNEY BLADDER
PITTA	{	3 Ether	LIVER GALLBLADDER
	{	4 Fire	HEART S. INTESTINE
KAPHA		5 Earth	STOMACH SPLEEN

Figs. 15a and 15b. The wheel of organ support and control in Ayurveda.
Reprinted with permission from *The Lost Secrets of Ayurvedic Acupuncture* by Dr. Frank Ros, Lotus Press, P.O. Box 325, Twin Lakes, WI 53181. © 1994 All Rights Reserved.

Nevertheless, Svoboda and Lade make some good points. Just to give one example, they point out that the efficacy of acupuncture can be greatly increased by modifying the usual technique according to the patient's dominant *dosha* imbalance. Historically, the Indian system does have something akin to acupuncture (see chapter 11), but it has never been fully developed. This is an area in which much cross-fertilization should benefit both systems.

Over time, there has been some cross-fertilization between the two systems in the area of therapeutic uses of plants, herbs, minerals, and so forth. These cross-fertilizations will continue, no doubt.

I also think that by combining what is theoretically the best in each system we may have an opportunity to develop a more complete vital body medicine. Suppose we adopt the Chinese version of the five elements theory and combine it with the threefold *guna/dosha* vital/physical body types of Ayurveda. We should then develop the predictions of such a model of vital body medicine and compare it with the facts of the clinical situation.

What is needed most is to bring back empirical research studies in both systems, traditional Chinese and Indian, especially now that we are moving toward a good theoretical understanding of both. As we research the vital body, perhaps one day we will even find a better substitute for the five elements theory that plays a role in both systems. Also, with research we may be able to correct one major omission in both systems: the chakras, which can be thought of as the feeling connections. In the next chapter we will look at the possibility of a vital body medicine based on the chakras.

Summary of the Important Ideas of Traditional Chinese Medicine

The following is a summary of the important ideas of this chapter, ideas for you to take in and consider for your own use:

• Traditional Chinese medicine is vital body medicine par excellence. Like Ayurveda, it also looks at disease as disharmony and imbalance of vital energy movements and tries to correct them via herbal medicine, an external infusion of *chi* (vital energy). Unlike Ayurveda, it also has the highly sophisticated and effective technique, acupuncture, which uses direct stimulation at the skin to correct vital movement gone awry.

• Chinese medicine is quantum medicine. It uses the twofold quantum wave (yang) and particle (yin) aspects of *chi* to classify the vital energy imbalance as well as the organ representations of the vital morphogenetic fields. In this way, Chinese medicine employs the wave-particle polarity of the underlying quantum dynamics of *chi* itself to classify the defects (akin to Ayurvedic *doshas*) of the vital body representations, the organs.

- Chinese medicine is also individualized medicine. A person may suffer from yang deficiency characterized by obesity and excess phlegm (akin to the Ayurvedic *dosha* of *kapha*). Or she may suffer from yin deficiency characterized by thin body, excess intestinal gas, and acidity (akin to the Ayurvedic *doshas* of *vata* and *pitta*).

- The basic self-help preventive principle of Chinese medicine is straightforward: Keep your yang (doing, movement) and yin (being, quiescence) in balance.

- The details of traditional Chinese medicine are developed combining the yin-yang classification with another classification based on the Chinese version of the five elements—wood, fire, earth, metal, and water.

- Acupuncture works not because the skin intrusion induced by acupuncture affects nerve signals, but because the skin puncture is able to influence vital energy movements. The puncture first affects the vital energy flow in the meridians correlated to the physical body at the skin, and second, through interior connections, the vital energy flow in the meridians inside the body connecting the vital blueprints of organs.

- There are many points of similarity between Ayurveda and traditional Chinese medicine, so much so that a two-way trade of techniques between the two systems seems highly useful.

---

1 1

Chakra Medicine

One other indirect proof of the authenticity of acupuncture and its meridians as a science is that Ayurveda also has these ingredients, although they are not employed as much in India. The founders of the Ayurvedic tradition independently discovered channels of flow of vital energy *(prana);* these channels they called *nadis.* They also discovered that certain points of intersection of these *nadis* are special and that physical stimulation at these points will connect to internal organs. These points are called *marma* in Sanskrit, and *marma* therapy is part of Ayurvedic massage therapy even today.

Ayurvedic practitioners even developed a sort of acupuncture—the art of piercing the skin with needles at the *marma* points—but it was not as sophisticated as the Chinese system, partly because the Ayurvedic practitioner's landscape for the vital body is not organ-specific. Further, in India, Ayurveda became part of a spiritual system called *tantra* and the entire focus of vital energy research became spiritual upliftment rather than the cure of disease. This is the subject of a later chapter.

In Ayurveda, instead of specific organs, the subject of *nadis* revolves around points of the body called the chakras (the Sanskrit word *chakra* means "wheel"). These chakras are fairly well known today, and one of the successes of the new integrative paradigm is that it enables us to understand the chakras scientifically. Once we do that, the possibility of developing new healing techniques using the chakras opens up—the subject of this chapter.

Like their Chinese counterparts, the Ayurvedic practitioners were also experts in diagnosis through pulse testing (which is testing the *nadis*). I grew up hearing many fantastic stories about the efficacy of diagnosis through *nadi* testing. I will tell you one of these stories.

An Ayurvedic physician was called by a Moslem king to examine the health of his wife and to suggest a suitable diet, except that it was a trick perpetrated by the king's courtesans. Married Moslem women cannot be seen or touched by other men (family excepted). So custom demanded that the woman would sit behind a curtain, a rope would be tied to her wrist, and the physician would get to examine only this rope (that is, read her *nadi* through the rope, a chore similar to trying to read her pulse through the rope).

The trick played on the "poor" Ayurvedic doctor was to substitute a cow for the wife. It is said that the Ayurvedic master examined the rope for a long time, apparently trying to read the *nadi,* and then said with a sigh, "I can't understand this. But all this patient needs is an ample diet of grass and that will take care of her."

What Are the Chakras?

I mentioned earlier that when we experience emotion, there is not only a physical effect and a mental thought but also a feeling that accompanies it. What do we feel? We feel the movement of vital energy accompanying the emotion. But where in the body do we feel our emotions? Or, putting it more accurately, where in the body do we feel the feeling component of our emotions?

If you are a connoisseur of feeling, your answer will be that of course that depends on the emotion and who you are. If you are an intellectual, it is likely that you only feel vital energy on the top of your head. When we are being intellectual, that is where the vital energy goes. This is the crown chakra (see figure 16).

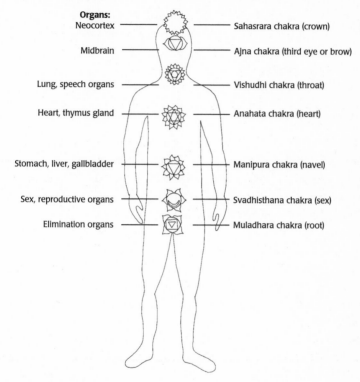

Organs:

Neocortex	Sahasrara chakra (crown)
Midbrain	Ajna chakra (third eye or brow)
Lung, speech organs	Vishudhi chakra (throat)
Heart, thymus gland	Anahata chakra (heart)
Stomach, liver, gallbladder	Manipura chakra (navel)
Sex, reproductive organs	Svadhisthana chakra (sex)
Elimination organs	Muladhara chakra (root)

Fig. 16. The chakras.

If you are not predominantly an intellectual, then you will recognize other places in the body where you feel your energy. The most familiar of these places is the heart chakra, the place where you feel romantic energy. Can you remember the first time you realized you were in love? Close your eyes and imagine that moment right now; soon you will feel the surge of energy in your heart chakra (felt as a throb, tingle, warmth, or expansion). This is why people read romance novels or watch mushy "heartwarming" movies. They like the surge of energy in their heart chakra (which is felt as warmth on these occasions).

In contrast, when we watch sex and violence on TV, the energy goes to the lower chakras and we feel more grounded. No doubt this is the reason for the popularity of sex and violence in the media today. People get tired of being in their head, which is where the demands of their jobs routinely take them.

When we feel good about ourselves, we feel an energy boost in our navel chakra; if we feel insecure, we feel energy going out of that chakra, such as "butterflies" in the stomach. We feel rooted when the energy moves into the root chakra, but when the energy drains out of there, we feel fear. The sex chakra is where the energy goes when we feel amorous.

After sex or a good meal, the energy rises to the heart chakra. People in earlier days seem to have known this. In those days, males handled the family finance, and women knew when to ask their male partner for household needs, especially money: after sex or after feeding him a good meal. It is all forgotten now, thanks to women's liberation, but in those days it was quite common to remark, "The way to a man's heart is through his stomach."

When we are nervous about giving a speech, our throat seems to dry up; this is because vital energy has moved out of the throat chakra. On the other hand, if you are communicating well, feel your throat chakra. You will enjoy the inflow of energy there; we all do.

When we concentrate, our eyebrows focus and you can feel heat in the point of your "third eye." This is also the chakra that "opens" when we have intuitive insights. In India, when people do spiritual work, a time for a great many intuitive experiences, the third eye becomes so hot that people put sandalwood paste there to soothe it. You may have seen Indian women wear a *bindi* (red dot) on their forehead; the reason is the same, at least traditionally (now Indian women may do it because it is stylish).

In summary, chakras are places in our physical bodies where we feel vital energy localized when we are experiencing a feeling.

The Science of the Chakras

If you examine figure 16, you will notice that each of the chakras is situated near one or more important organs of our body.

This has been noted for millennia, and this is the clue suggesting a scientific understanding of chakras.

Remember Rupert Sheldrake's work on morphogenetic fields? Sheldrake correctly theorized the function of the vital body: It provides the blueprint for physical form-making. Physical organs are the representations of vital body functions such as maintenance of the body or reproduction. Chakras are those places in the physical body where consciousness simultaneously collapses the vital and the physical in the process of which the representation of the former is made in the latter.

Here is a chakra-by-chakra description of the vital function, the corresponding physical organs, and the associated emotional feelings:

Root chakra: The vital body function is elimination, a crucial component of maintenance called catabolism. The organs that express the function are the kidneys, bladder, and large intestine (rectum and anus). The feelings are selfish rootedness and survival-oriented competitiveness when energy moves in, and fear when the energy moves out.

Sex chakra: The vital body function is reproduction. The reproductive organs—uterus, ovaries, prostate, and testes—are the physical representation of the reproductive function. The feelings are sexuality and amour when energy is inward and increasing; when energy is outgoing and depleting, the feeling is of unfulfilled lust.

Navel chakra: The vital body function is maintenance (anabolism), and the organ representations are the stomach and small intestine, liver, gallbladder, and pancreas. With an upsurge of energy at this chakra, the feelings are pride and anger; when energy moves out, the feelings are unworthiness and resentment.

Heart chakra: The vital body function is self-distinction (the distinction between me and not-me). The organ representations are the heart and thymus gland of the immune system whose job is to distinguish "me" from "not me." Here we feel romance when energy moves in. When energy moves out, we feel loss, grief, hurt, and jealousy.

Why is romance felt at the heart chakra when we meet the

appropriate partner? Because now the "me" is extended to include the partner. But romantic love is still me-oriented; he/she is important because he/she is mine. This is to be expected. These feelings are the conditioned movement of vital energy, conditioned through millions of years of evolution. When the self-distinction extends to everyone, when everyone is my "family," the heart chakra is said to "open" for everyone and we feel the universal, unconditional love that mystics talk about as agape or compassion.

Throat chakra: The vital function is self-expression. The organ representations are the lungs, throat and speech organs, hearing organs, and thyroid gland. The associated feelings are the exultation of freedom (of speech) when energy moves in and frustration when the opposite happens. (You recognize why freedom of speech is considered so important in our culture, although real freedom is freedom of choice.)

Brow chakra (also called the third eye): The vital function is evolution. The evolutionary impulse from the supramental for the development of the neocortex, the physical representation of the mind, was "heard" here via intuition. Hence this is also the chakra of intuitive energy. The organ representations are the organs of the middle and the hind brain, the eyes, and the pituitary gland. Next time you focus on a problem, notice the vital energy enhancement at this chakra. The associated feelings are clarity of understanding (when energy moves in) and confusion (with depletion of energy).

Crown chakra: The vital function is self-knowledge, for which the organ representation is the neocortex where the mind is mapped, mind that transcends the vital body. The gland here is the pineal gland. The associated feelings are satisfaction (when energy is gained) and despair (when energy is depleted).

Of crucial significance is the fact that there is an endocrine gland associated with each of the chakras. The endocrine glands communicate with the brain where the mind is mapped. In this way, through this psychoneuroimmunological connection as well as through the autonomic nervous system, the mind gets control over the vital energies (see chapter 14).

As mentioned before, the emphasis of Ayurveda is healing,

whereas the emphasis of tantra, with which the Ayurvedic ideas of chakras and *nadis* became entangled, is spiritual awakening. Thus the writings on chakras have become highly confused between these two. For example, people carelessly talk about the "opening" of a chakra through simple techniques of massage. In a way, the language is not wrong. There can be disease at one of the organs of a chakra if the conditioned movement of the correlated *prana* has become stagnant or blocked somehow. Massage certainly can clear up the stagnancy, and this can be called an "opening." But this is not the opening that tantra talks about.

Tantra is about creatively transforming the conditioned movement of *prana* altogether. The conditioned movements of *prana* at the chakras, together with conditioned movements of the mind that we call our habit patterns, comprise our ego-character. This character is a persona, a very self-oriented one; it's a mask that hides our true self—the quantum self.

Along with the conditioned movement of *prana* associated with the conditioned feelings stated previously, at each chakra there is a potential movement. We realize this potential through creativity. I have already mentioned how possessive romantic love has the potential to transform into universal love. Similarly, when we creatively transform the energy of the root chakra, insecurity transforms into confidence and fear into courage (courage like that shown by Gandhi comes from this transformation); at the sex chakra, sexuality and associated negative feelings of lust have the potential to transform into genuine respect for self and others; at the navel chakra, pride and unworthiness have the potential to transform into worthiness and true self-worth.

Among the upper chakras, at the third eye or brow, clarity and confusion of intellectual understanding can become intuitive understanding when transformed; and at the crown, satisfaction and despair transform into permanent easy-without-effort happiness (*ananda* in Sanskrit). Energy no longer gets depleted.

The goal of tantra is to transcend the ego-character to wake up to the universal self, which I call the quantum self. When this transcendence takes place at the vital level, the conditioned movement is redirected through a new channel, and the chakras "open up" along the pathway of this new channel. This is wholeness, the meaning of "holy," and the ultimate goal of healing.

Chakra Medicine

Meanwhile, leaving aside the lofty goals of tantra for now, a chakra medicine is gradually developing that looks for disease and healing in terms of the nonnormal movement of *prana* and its correction at each of the chakras. Such nonnormal movement may consist of the excessive movement of energy at a chakra or an unnatural lack thereof, and also stagnancy or blockage.

For example, if the energy moves out excessively from the root chakra, via the stimulation of fear, at the physical level there would be too much secretion from the adrenal gland (flight-or-fight response), which, if happening frequently, can contribute to disease such as chronic fatigue syndrome. As will be discussed in a later chapter, mind also gets into the game through the psychoneuroimmunological connections.

Relevant here is the work of the physician Christina Page (1992), who unabashedly classifies many organ diseases as abnormal energy movement at one chakra or another. Here is a sample of diseases that Page lists (I have altered her list slightly) as possible if energy movement goes awry at the chakras:

Root chakra: constipation, piles, colitis, diarrhea.

Sex chakra: impotence, vaginismus (tightness of the vaginal muscle), prostate disease, disease of the female reproductive system.

Navel chakra: irritable bowel syndrome, diabetes, peptic and gastric ulcer, liver disease, hiatal hernia.

Heart chakra: heart disease, immune system disease, cancer. Cancer is listed as a heart chakra disease for good reasons. Cancer results because cancer cells are abnormal cells and, in principle, the immune system should be able to detect them and eliminate them. So cancer results perhaps because of an immune system malfunction (including the thymus), and therefore cancer may be related to abnormal movement of vital energy at the heart. Of course, once developed, cancer can spread to any and all organs;

then it also becomes related to the vital energy abnormality at all the relevant chakras.

Throat chakra: thyroid overactivity or underactivity, asthma, sore throat, ear disease.

Third eye: migraine and tension headaches, eye disease, sinusitis.

Crown chakra: epilepsy, Alzheimer's disease, mind-brain diseases such as depression and schizophrenia.

Chakra medicine consists of complementing the treatment of the physical symptoms (through allopathy) and *pranic* imbalance (through Ayurveda, Chinese medicine, or homeopathy) with psychological work on the mind-set and psychic healing through direct *pranic* infusion by a *pranic* healer to the affected chakra.

The good news is that almost everyone can be a *pranic* healer because we all have psychic ability to that extent. Although in the West, not too many people are familiar with feeling the movement of *prana,* it is easily learned. A simple exercise is to rub your palms together and then bring them apart by about half an inch in the East Indian style of greeting called "namaste" (which, by the way, means "I greet you from the place where you and I are one").

You will feel tingles (convince yourself that they are not blood circulation or nerve impulses), which is the movement of *prana* in the skin. You can amplify the tingling by outstretching your arms, opening your palms to the canopy of the sky, and inviting all the healing *chi* that the universe is willing to send you. Now your palms are energized and you are ready to give *pranic* healing to a friend.

Let the friend lie down comfortably and maintain a receptive mind during the exercise. Bring your energized palms (you may need to energize them more than once) close to each of your friend's chakras with the intention of healing the chakra; no physical touching is called for. Start with the crown chakra and go down until you finish with the root chakra. Maintain your awareness throughout.

I often let participants do this exercise in my workshops. I myself learned it from the mystic-physician Richard Moss, who calls it "sacred meditation," but now I find quite a few others are also teaching it in their workshops, which is a very good thing. It is about time that vital energy ceases to be something mysterious and esoteric that only Easterners can experience and use.

There are also simple practices we can do that help energy movement at the chakras. For the root chakra, walking barefoot on the ground (not cement) or working with soil, as in gardening, is helpful. For the sex chakra, tantric sex, that is, sex without emphasis on orgasm, or even just cuddling is helpful to eradicate objectification of the opposite sex. For the navel chakra, sensitivity training—being not so sensitive to other people's negative energy—is useful.

For the heart chakra, laughing meditation—laughing without stopping—is good; generally keeping in a happy frame of mind, reading inspiring books and love stories, even seeing "mushy" movies can help. The "Mother Teresa effect" has now become famous. Harvard students watched a documentary about Mother Teresa's work, and a subsequent blood test indicated that their immune systems had increased levels of an immunoglobulin as a result. Their immune system became stronger, albeit temporarily, because vital energy had moved into their heart chakra.

For the throat chakra, free expression of creativity during ordinary living—singing in the shower, reciting a poem, chanting—can clear up this chakra in a hurry.

For the brow chakra, there is a breathing practice in the yoga tradition called *kapalavati*, which means "radiant forehead." Sit comfortably, start with a small inhalation, and then practice forced exhalation of breath using only your stomach muscles for about 20 to 40 times a minute. If you are doing it properly, when you stop you will experience a few moments of breathlessness; don't hurry to inhale—your body will do it as soon as necessary. When you are without breath, notice that you are also without thought. In that thoughtlessness, intuitions may come to you.

Meditation and mindfulness in general are good practices to keep the crown chakra in dynamic balance.

Special emphasis must be given to breathing practices called *pranayama* in chakra cleansing practices. The movement of *prana* along *nadis* in the vital body parallels, and is correlated with, the movement of air as we breathe; this is the basis of *pranayama*. When we breathe shallowly, we feel the breath only in the nose and throat area. As we breathe deeper, we begin to feel the breath in the chest. When we breathe the deepest, we can feel the breath in the stomach area. One form of *pranayama* consists of just breathing deeply with awareness. If we do this practice, we soon become aware of all the vital energy movements along the *nadis* roughly along the spine and embracing the major chakras.

As our breath slows down, and the practice of *pranayama* will do that to the breath, this movement of *prana* along the *nadis* connecting the major chakras also slows down. This has the effect of increasing our awareness of the movement of *prana;* in particular, we become aware of the gap between the quantum collapses of the *pranic* movement. In these gaps, there is unconscious processing, and quantum possibilities of *pranic* movement can proliferate. So we have better and better access to collapsing new movements of *prana,* better and better chance for creative quantum leaps. The more quantum leaps, the more there is balancing of previously unbalanced patterns of *pranic* movements (which includes the balancing of the chakras).

As mentioned earlier, chakra medicine has given us a chakra psychology in which a therapist tries to correct the mind-set that is causing the energy imbalance at the chakra. Sometimes this is appropriate. For example, breast cancer may indicate lack of self-love, and psychotherapy, which can help rekindle self-love, may be helpful. But one can also overdo chakra psychology (see also chapter 15).

Chakra medicine may also be connected with astrology. The idea is simple. The archetypes that set contexts of the vital functions of the vital morphogenetic fields may also be the "guiding angels" for setting up how your astrological horoscope relates to the movements of the solar planets, the moon, and the Sun. This gives you additional astrological guidelines of health maintenance, even healing.

Although astrology is regarded as "bah humbug" in conventional materialist science, the idea that outer and inner should be connected makes a lot of sense when we engage in primacy-of-consciousness science. Look for much progress in our understanding of the place of astrology in our healing science in the future.

What to Take Away from This Chapter

- The major theme of this chapter is the developing science of the chakras, points in the physical body where we feel our feelings. If you are unfamiliar with the movements of vital energy at the chakras (which give you feelings), I strongly encourage you to familiarize yourself with these movements whenever you are in the swirl of emotions.

- I also encourage you to do some practices as suggested in the book (such as the one called *pranayama*), which are like meditations; instead of thoughts, you are using vital energy movements. These practices will help you balance creativity and conditioning in your processing of your vital body.

- Chakra medicine is based on the idea that diseases of our major organs can be due to the imbalance or blocks of vital energy movements at the corresponding chakra.

- You can heal yourself by regaining balanced movement of vital energy at the chakra, or through unblocking the movement that was suppressed (which means making what was previously unconscious conscious).

12

Is Homeopathy for Real?

I still remember one of my childhood experiences of the miracle of homeopathy. I was a 12-year-old kid, popular, active in sports, good in academics, but I was hugely unhappy because I had an utterly embarrassing thing happening to my body—warts. And they grew everywhere on my body. We tried various wart-removing agents; nothing worked. Finally somebody suggested homeopathy. I still remember the medicine I got, Thuja 30x—four little white globules that tasted sweet. I had to suck on them until they dissolved in my mouth. After two days, one by one the warts just fell off my body or something like that. I was cured. I was relieved. It was miracle medicine.

Actually, at the time I didn't appreciate the entirety of the miracle that homeopathy is. I did not know that calculations can easily prove that on the average it was unlikely that there was any medicinal substance at all in the dilution of Thuja that I took, in the conventional sense of what we mean by "substance." In the conventional way of thinking, four little sugar pills cured my disease.

Such a "sugar pill cure" under the guise of medicine so the patient thinks he is getting something "good" from a qualified healer is today called a placebo cure. When told a story like mine, most allopaths dismiss homeopathy, saying it is just placebo. What complicates the issue is that a disease like warts has also been known to be cured by placebo (Weil 1983).

Of course, a placebo cure seems to be a miracle. What we know about the immune system and the body's defense mechanism suggests something. Disease happens to us when the immune system response, the body's defense, is not working properly. A placebo works by raising the mind's *expectation* of healing, which then triggers the body's defense mechanism to work properly once again. But what does the triggering? Four little sugar pills? That would be an even bigger miracle than homeopathy! The mind? But from the allopathic point of view that is a miracle too, because it suggests mind over matter.

Is homeopathy a placebo? Many studies have been done and the outcome is still controversial, although I have read that a few studies have been definitive (see Pelletier 2000). So let's ask the question in another way. Does homeopathy work? Can it work in the face of the allopath's correct criticism that you are not administering even one molecule of the medicinal substance in some of the dosages? If it works anyway, how does it work?

In what follows, I will give a plausible theory for homeopathy (for a good introduction to the existing literature, read Vithoulkas 1980, Ullman 1988). Such a plausible theory can be constructed easily once we accept the vital body and the vital body's connection to the physical—namely, that the physical organs are but representations of the vital body morphogenetic fields.

Now I should be a little more systematic and should tell you the two fundamental axioms of homeopathy discovered by its founder Samuel Hahnemann. The first axiom is "like cures like" (*similia similibus curentur*, in Latin). If a certain medicinal substance produces a certain confluence of symptoms in a healthy body, then that substance will act as a cure to a diseased person with the same symptoms when given in a potentized (highly dilute) homeopathic form.

An example, for which I choose Hahnemann's very first case study of like cures like, will make this clear. Hahnemann, while translating a book by William Cullen, became aware that Peruvian bark or cinchona can be used to treat malaria because cinchona is "bitter." This bitterness as an explanation seemed absurd to Hahnemann, so he began experimenting with cinchona himself, taking small doses of the substance although he was a well person:

> I took by way of experiment twice a day, four drach-mas of good China cinchona. My feet, finger ends, etc., at first became cold; I grew languid and drowsy; then my heart began to palpitate, and my pulse grew hard and small; intolerable anxiety, trembling, prostration through-out all my limbs; then pulsation in the head, redness of my cheeks, thirst, and, in short, all these symptoms which are ordinarily characteristic of intermittent fever, made their appearance, one after the other, yet without the peculiar, chilly, shivering rigor.
>
> Briefly, even though symptoms which are a regular occurrence and especially characteristic—as the stupidity of the mind, the kind of rigidity of the limbs, but, above all, the numb, disagreeable sensation, which seems to have its seat in the periosteum, over every bone in the body— all these made their appearance. This paroxysm lasted two or three hours each time, and recurred if I repeated this dose, not otherwise; I discontinued it, and was in good health (quoted in Grossman 1985, p. 60).

Even today, homeopathic researchers do their "provings" more or less the way Hahnemann did it that first time. Take a sub-stance that you have some reason to believe to be medicinal, and administer it in small doses to well subjects who will note all their developing symptoms. These empirical provings then become part of a *Materia Medica,* which physicians can consult to find the like-cures-like medicine for a disease through similarity of symptoms.

From an allopathic point of view, this like-cures-like axiom is not all that objectionable. Hippocrates wrote, "Through the like, disease is produced; and through the application of the like, disease is cured." Allopathic medicine uses vaccines that are based on a similar idea. But allopathic practitioners don't believe, cannot believe, this like-cures-like principle to be universally true. And there the allopathic doctor is justified because there are everyday examples of violation of this rule of cure in allopathy; for example, aspirin cures a headache, but it is actually antipathic—"unlike cures like"—in terms of symptoms.

The second axiom of homeopathy is the one that aggravates the materialist allopath the most. This is called the "less is more" axiom: the more you dilute the medicinal substance in water (using a certain procedure, described as follows), the more potent the effect is. And as I said before, the dilutions homeopaths routinely prescribe for their patients are very high, materially speaking.

Consider: One part of a medicinal substance is diluted with nine parts of an alcohol-water mixture. This mixture is thoroughly shaken ("succussed," to use the technical term) about 40 times; after this, nine parts are discarded, and the remaining one part is diluted once again with the water-alcohol mixture. This is "succussed" once again, and the process of dilution and succussing may be continued indefinitely, producing medicine of increasing potency denoted as 1x, 3x, 6x, 30x, 200x, and so forth.

Allopathic physicians ridicule homeopathy because after a certain dilution it is mathematically extremely unlikely that even one material molecule of the "medicine" will be present.

Let's be a little technical to drive in the allopath's seemingly valid point. There is a law, Avogadro's law of chemistry, that says that a "mole" (which stands for the equivalent of the molecular weight of a substance in grams) of any substance contains on the order of 10^{24} molecules of the substance. So after a homeopathic dilution of 24x of a mole of a medicinal substance (which chemically means a dilution by a factor of 10^{-24}), it is not likely that there will be a single molecule of the medicinal substance present.

But both homeopathic axioms make sense from a vital body point of view. From the vital body perspective, the immune system is a physical representation of the vital blueprint for defending the body against external or internal antigens. A disease means that this physical representation, in its current form, is not working properly. It makes sense that this malfunction may be due to the malfunction of the vital blueprint (an imbalance of the relevant movements of vital energies) itself. If that is so, we have to infuse the proper vital energy to fix the vital blueprint in the system to effect a cure.

But what is a "proper vital energy" for treating the vital body? Like cures like. If the medicinal substance incited the same symptoms in a healthy body as the symptoms of the diseased body, it must mean that the vital energy movements of the medicinal substance and the body's relevant vital energy movements in this case (which are in imbalance) resonate in a certain sense. The vital energy movements of the medicinal substance can then be used to balance the imbalance of the relevant vital energy movements of the diseased person.

The principle "less is more" also makes sense. The medicinal substances are often toxic at the physical level. If the cure is in the vital plane, the physical is irrelevant, and what is the wisdom of administering unnecessary toxic stuff to the organism? So dilution makes sense: Eliminate the "physical body" of the medicine while the vital aspect of it remains intact.

As mentioned, homeopaths use a very elaborate procedure of shaking the mixture before making further dilutions. Even that now becomes plausible. Perhaps this procedure makes sure that the vital energy, correlated before with the physical part of the medicine alone, now becomes correlated with the entire alcohol-water mixture. (How? Perhaps through the conscious intention of the preparer, all that movement helping to trigger that intention!)

That conscious intention may be involved is further indicated by the ability of some psychic homeopaths to "administer" the remedy by thought, intending that the pattern of vital energy of the appropriate plant (the source of the homeopathic remedy) be imprinted on the vital body of the patient. What is amazing is patients are healed this way, proving the power of downward causation even when applied nonlocally.

Homeopathy adds an interesting dimension to vital body healing. Not only the "friendly" herbs of Ayurveda and traditional Chinese medicine can be used, but also the often "unfriendly" (that is, harmful) substances of allopathic medicine can be used for healing.

One intuition that homeopathy opens up for me is the question of homeopathic food: Can we not make concentrates of vital energy essence of food by the homeopathic procedure of dilution and use them as food supplements? This may be useful because today with fast-food preparation, refrigeration, freezing, and the addition of preservatives, I am not sure we are getting good nutrition at the vital level, in which case homeopathic food may be a solution.

Believe it or not, there are some data in support of this idea. Stanley Rice and his collaborators used microdoses of a fertilizer to enhance crop production, this in the middle of the 1970s' oil crisis. In homeopathic terms, the dilution could be called 9x, but even so, the published results showed that tomato and sweet corn plants treated this way produced 30 percent and 25 percent more fruit, respectively.

And here's another point. I have noticed that protein supplements that vegetarians often use to achieve proper protein balance in their diet don't quite do the job of protein-containing foods such as tofu. I am convinced that this is because the protein supplements, in the process of their extraction from the original protein-containing food, lose all the vital energy correlated with the food. So to get the complete effect, we have to supplement the supplement, and here a homeopathic rendition of the associated vital energy of the food from which protein supplements are extracted would be wonderful.

The Question of Individuality

In summary, homeopathy is not all that different from allopathy once we recognize that it is a medicinal system to cure the vital body imbalances in the same way that allopathy tries to cure physical body imbalances. In fact, with respect to how to search for a suitable medicine for a certain ailment, allopathic methods are more of a shot in the dark than the like-cures-like philosophy of homeopathy.

There is another aspect of homeopathy that is similar in spirit to Ayurveda (and traditional Chinese medicine). Homeopathic medicine is also used for ailments of a constitutional nature, ailments that people tend to accept as part of their physical makeup or constitution.

For example, many people have a tendency for obesity from childhood, and they learn to accept it. As they approach middle age, obesity can become more of a health problem than before. Rationalization does not help; dieting does not help. So is there any way of losing weight aside from the drastic allopathic surgery of cutting out part of the small intestine? If you are such a person and you go to a homeopath, she may ask, "Are your hands limp?" and pick up your hands and examine them. The doctor may also ask if you are "fearful in nature." If you have all three symptoms, there is hope for you. The doctor will prescribe *Calcarea Carb*. And voila! With hardly any change in diet and without a huge regimen of exercises, you may start shedding weight (Ballentine 1999).

The base medicine of *Calcarea Carb* is limestone, the same as the sedimentary rock that makes up much of the Earth's crust. As you undoubtedly know, sedimentary rocks have a lot of history with the living part of the planet, lots of correlation with vital energy. And perhaps the three symptoms above correspond to a vital energy imbalance that can be fixed with the vital energy that comes with *Calcarea Carb!* Once the vital body, the morphogenetic fields, have been "fixed," effortlessly, the physical body organs, the representations of the morphogenetic fields, will be corrected also and will function normally without giving rise to the cravings that make dieting impossible for such obese people.

Notice, however, that a *Calcarea Carb* patient may sound like a *kapha* person, but there are striking differences. In Ayurveda, a person with *kapha prakriti* will in general have a tendency to be robust, but that person will be also stable and happy by nature, not fearful. So *kapha prakriti,* so long as it is *prakriti,* natural homeostasis, is okay. Only when we have an imbalance that takes the person away from the natural homeostasis of *prakriti* to a lack of it is there an ailment, a disease that needs to be treated. The symptoms of clamminess (shutting up like a clam) and fearfulness suggest the deviation from the natural *kapha prakriti* and then homeopathy comes as a handy treatment.

In homeopathy, individuality is important and all symptoms are important. It was recognized from the beginning that causes of disease may be subtle, undetected by the quantitative measurements of physical tools, and therefore symptoms are the only clue to them.

Allopathic physicians of the nineteenth century believed, as do conventional doctors even today (Coulter 1973; Grossman 1985), that:

1. Diseases are events that have determinable causes.

2. The classification of diseases should be according to their causes.

3. Symptoms indicate their causes. Thus those symptoms that are directly related to the causes are more important for treatment purposes than those only distantly related to the causes.

These Hahnemann countered with his homeopathic claims:

1. Disease is a breakdown of vital forces.

2. The internal cause of the breakdown cannot be known.

3. Disease cannot be classified according to the internal cause.

4. Diseases can be known only according to their symptoms; hence all symptoms are of equal importance.

When homeopathy is integrated with Ayurveda and Chinese medicine, and our understanding of the vital body grows, I believe that a classification of vital body diseases according to vital body "internal" causes may be available to us. Of course, due to the quantum dynamics of the vital body, the diagnosis will always require a considerable amount of intuition. The development of such intuitive diagnostic tools is, in fact, the big challenge of vital body medicine.

But until this happens, the homeopathic strategy of a thorough analysis of all symptoms, not just the obvious ones that are causing suffering, is a good one. This strategy helps to identify the vital energy *signature* of the particular medicinal substance that will resonate the best with the unbalanced vital energy *signature* of the individual patient.

The important point here is that our vital body is individualized through the specific history of our conditioning (maybe even for many incarnations). When the vital movements go awry, causing disease at the physical level, there must likewise be an individual signature to the vital energy imbalance.

I will tell you my own experience with the kinds of personal questions that homeopathic doctors sometimes ask to make their diagnosis. I had a certain ailment (I don't remember its nature after all these years) that was bothering me a lot. In my interview with the homeopath, one question I remember is "When you walk, do your feet hit each other often, although it is not your intention?" And this was indeed the case, although I would not in a million years have made the connection of that symptom with my disease. Anyway, the good doctor gave me a medicine and it worked.

Hering's Law of Cure

Medicine is a difficult science because unlike physics and chemistry the cause-effect relationships seem to be more subtle. For example, when cold viruses are around in a certain environment, some people will get an infection, but others won't. The same is true of bacterial infections. Some easily contract the disease, but others don't.

Allopaths generally attribute this kind of difference from person to person to the immunity the person has, which is, of course, an idea that is not quantifiable, not even verifiable. Homeopaths, starting with Hahnemann, and especially emphasized by Constantine

Hering, who is sometimes referred to as the father of American homeopathy, take a different tack. To homeopaths, disease has a vital body origin.

What we call immunity is really a measure of the balance with which the vital energies are functioning. If this balance goes awry, then there should be symptoms of that imbalance that are precursors of the aggravated symptoms that occur at the stage when we clearly see that there is disease.

Allopathy treats only the later aggravated symptoms of the physical body. Once this happens, the earlier symptoms, homeopaths insist, recede to the background giving one a false sense of wellness. So classical homeopaths advise against concurrent allopathic treatment with homeopathic treatment. With homeopathic treatment, the aggravated symptoms will certainly go away also, albeit a little slower. But then comes the big bonus of homeopathy over allopathy: After the aggravated symptoms disappear, the earlier symptoms will also begin to disappear. Symptoms disappear in the reverse order in which they appear. This is referred to as Hering's law of cure.

Is this law justified? Hering gave the reason why it is justified, which remains essentially correct, even for our new model of science within consciousness: Healing under homeopathic treatment begins with the most "profound" part of the organism and moves to the more superficial parts. The word "profound" now has to be interpreted to mean the vital body and "superficial" to mean the physical body. Since the superficial parts—the representations—are quicker to heal than the vital imbalances, the superficial symptoms disappear before the profound symptoms.

Is there any basis to the allegation that allopathic treatment may harm the patient? Do the more profound symptoms actually recede with an allopathic cure of the superficial symptoms? The answer to both questions is yes. An allopathic treatment can cure physical, superficial symptoms, but only at the expense of great harm to the physical body, including its representation-making apparatus. So the physical body will be unable to make proper representations of the ongoing imbalances of the vital body, leading to a chronic mismatch between the vital body blueprints and their physical body representations. Such a mismatch is what we feel as pain, which is part and parcel of chronic diseases (see also chapter 14).

Classical versus Modern Homeopathy

I will end this chapter with a brief mention of "modern" homeopathy—a product of the French school. When homeopathy was officially co-opted in France, the grateful homeopaths (like the modern Ayurvedic physicians) made a large number of philosophical concessions, the major one being to stop all talk of a "vital force." It was also considered okay in modern homeopathy to have simultaneous allopathic and homeopathic treatments. Further, the idea of one homeopathic medicine for one unique individual with a disease was abandoned in favor of offering a combination of a few remedies. These changes made homeopathy a lot like allopathy, thus reducing the resistance of allopathic practitioners.

I hope with a proper science of homeopathy, of which you are witnessing a beginning here, the apologetic modern version of homeopathy will lose its appeal. Homeopathy is vital body medicine; otherwise it is nothing and does not make sense. This should be clear to everybody. But with the importance of the vital body recognized within science, especially in the metaphysics of medicine, homeopathy can return to its previous glory. There is no doubt that allopathy is invasive; it does harm even in simple treatments where its efficacy is clear, as in the case of a bacterial infection treated by antibiotics. If we could avoid the invasive procedures of allopathy, and this is possible in all cases where there is no life-threatening urgency, and instead embrace the gentle but more fundamental cure of homeopathy, our healing practices would be better for it.

A physician character in one of Moliere's plays said, "It is better to die according to the rules than to recover against the rules." I hope I have demonstrated enough plausibility for a scientific model of homeopathy to suggest that homeopathy is not "against the rules." Perhaps now the allopathy aficionados can relax. Of course, we need better ways than we have now to diagnose the right cause and the right treatment, and there is much to be done to perfect homeopathy, but if fundamentals are understood as they seem to be, can the details be far behind?

Conclusions

I have reached the following conclusions about homeopathy; I hope you will agree with me.

- Homeopathy is vital body medicine. If you don't subscribe to the existence of the vital body, homeopathy and its "less is more" philosophy will only baffle you. If you accept the vital body, not only will that enable you to understand why less is more, but you will also marvel at the intelligence of homeopathy as a medicine system.

- Have no doubt, homeopathy is a quantum medicine. The quantum principle of nonlocal correlation is essential to how homeopathic medicine is prepared and how it is administered.

- Homeopathy, like Ayurveda and traditional Chinese medicine, is also an individualized medicine, with one big difference in operational philosophy. Unlike Ayurveda and Chinese medicine, homeopathy, following its founder Hahnemann, has chosen to remain strictly empirical, steadfast in the belief that disease cannot be classified according to internal causes and can be known only through the symptoms.

- Some thinkers (see, for example, Coulter 1973) believe that this strict adherence to empiricism is a virtue and homeopathy is the most scientific of all medical systems because it is strictly empirical.

- But as Einstein said to Heisenberg, what we see depends on the theories we use to interpret our observations. Strict empiricism is a mirage, and one must try to develop theory in order to do science. I believe that as we gain experience with the vital body, we will begin to understand this question of individuality better than we do at present in Ayurveda and traditional Chinese medicine. Then some of homeopathy's spectacular success in finding individual cures on the basis of what is now considered "unusual" symptoms will make sense theoretically as well.

PART 3

Mind-Body Medicine

1 3

Quantum Mind, Meaning, and Medicine

Mind-body medicine does not make sense until you realize that it is not a consequence of mind over body, but a consequence of consciousness over body. Both body and mind are quantum possibilities of consciousness.

In an event of collapse of the waves of possibility, consciousness uses the mind to give meaning to some of the physical actualities collapsed. A part of the collapsed physical actualities (the brain) also makes a representation of the mental meaning. If the meaning that mind gives for an otherwise meaning-neutral stimulus is disharmonious, taking you away from ease, watch out. Disease may result. But consciousness, too, has the capacity to change the mental meaning, so healing is within its purview as well. This is called mind-body healing.

This is a summary of what's to come in the remainder of this book. The details are quite fascinating. For example, let's consider mind's quantum nature. Is there any evidence that the mind is a quantum body? Yes. For evidence, look at the modes of movement of the mind—thoughts.

The physicist David Bohm, who wrote one of the first books on quantum mechanics in the post–World War II era, needed a simple example of the quantum uncertainty principle. Recall that the uncertainty principle can be stated as follows: For a material quantum object, we can simultaneously measure *either* the position or the momentum with utmost accuracy, but not both. The example Bohm came up with can be used also as an uncertainty principle for the movement of thought: For thought, we can determine either the content (feature) or the direction of thought (association). Centering on one irrevocably affects the measurement of the other. Try it and see.

If you center on the feature (for example, when you recite the same thing in your mind like a mantra), you lose the direction of where your thought was going. If you free-associate, you won't be able to recall the contents of your thought later. Clearly, psychiatrists should know about the uncertainty principle for thoughts.

So for thoughts, feature and association are complementary variables satisfying an uncertainty principle. This suggests a quantum dynamics for the origin of thought.

Why is the physical world experienced as external and the mental world experienced as internal? There is no more dramatic difference than this between the mind and the physical. We now can understand this difference on the basis of the quantum nature of the mind. The argument is the same as I gave for the vital body (see chapter 8). Let's recap.

The crucial point here is to recognize, as Descartes did, that the physical body is *res extensa,* body with extension, and therefore a body subject to division. In other words, the physical world is distinguished by the division of micro and macro, the latter being conglomerates of the micro.

In the physical world, we do not have direct conscious access to the micro. We see the micro only with the amplifying help of the macro, the measuring-aid apparatuses. But there is a reward. Once the measurement is made and a particular pointer reading of the measurement apparatus has been chosen out of a myriad macro possibilities, the pointer does not run away, jumping on the train of quantum uncertainty. Its possibility waves are very sluggish, almost to the point of certainty, a certainty that can be shared by many observers. As a result, physical objects are experienced as parts of a shared reality, an external reality in awareness.

But mind, *res cogitans,* is without extension, it is one thing. It is like the infinite medium of the physicist in which there can be waves, and thoughts are such waves. However, there is no micro/macro distinction in the mental world. So we experience thoughts directly without the intermediary of amplifying apparatuses, but we pay a price. One price is that one person's experiencing a thought object affects the thought object due to the uncertainty principle, so that it is impossible (normally) for another person to experience the same thought object in an identical manner. Thoughts are private, thus experienced in awareness as internal.

The other price for the lack of micro/macro distinction in the realm of thought is that it is impossible to develop a tangled-hierarchical quantum-measurement apparatus. So mind can exist independent of the brain, but its movements can be registered and experienced in consciousness only when correlated with a physical brain.

Further evidence for the quantum nature of thought is exhibited in mental telepathy in which two (nearly) identical thoughts are simultaneously collapsed in two locally separated correlated subjects without any local connection between them. Distant viewing experiments, in which, for example, one psychic looks at a statue in the town square and a correlated psychic draws a picture of the statue while sitting inside a closet, have become famous.

A very important proof of the quantum nature of thought is discontinuity as exhibited in the phenomenon of creativity. Since creativity is an important aspect of healing, let's consider creativity in some detail.

Quantum Creativity

What is creativity? If you seriously examine this question, you realize quickly that creativity is the discovery of something new of value. But one problem will puzzle you: How do you define new? The new in creativity refers to either new meaning in a new context or new meaning in an old context or combination of contexts. Notice the difference between context and meaning: Context defines the meaning of words in such sentences as "You are blind if you don't see that meaning resides outside matter" or "Helen Keller was blind and could not see."

Creativity requires consciousness, or else who perceives the new meaning or value of the creative product (Goswami 1996)?

Can matter process meaning? Look at Picasso's *Guernica* or any other painting. From a material point of view, all you see are canvas, paint, molecules, and such—no meaning anywhere. You ascribe meaning to the picture with your mind, as the biologist Roger Sperry (1983) loved to point out to the physicalists.

You can understand this through the analogy of a computer. Analogically, computer scientists equate the brain with the hardware of a computer and the mind with the software. There is no need to propose a separate mind since for a computer we can easily incorporate the software in the hardware itself—then hardware is the software. But there is a crucial flaw in this "hardware is software, brain is mind" thinking.

As discussed in a previous chapter, computers are symbol-processing machines, nothing more than symbols acting on symbols. The semantic content of the symbols, the meaning of the software operations, lies in the mind of the programmer. You may think, why don't we reserve some symbols to tell us the meaning of symbols? But then you will need more symbols to tell you the meaning of meaning symbols, and so on, *ad infinitum*. You are seeing a consequence of the theorem discovered by the great mathematician Kurt Goedel: All sufficiently elaborate mathematical systems are incomplete if they claim to be logically consistent (Banerji 1994).

So to process meaning requires mind in addition to matter, which merely represents the mental meaning. A sculptor's hands process plaster of Paris, but what guides her action is a mental picture of what she wants to sculpt.

So when is our attempt at a sculpture, a painting, a musical score, an act of exploration in science creative? Ordinarily, thoughts conform to a conditioned pattern, a pattern that algorithmically follows in a step-by-step, continuous way from the contexts we know, giving us the similarity of thinking and computation. But such step-by-step thinking will not lead to a creative act, the discovery of new meaning in a new context.

Archimedes was told to verify that a gold crown was made of real gold. The mass he could measure; if he could calculate the volume, then dividing the mass by the volume would yield a quantity called density for which gold has a unique value. Archimedes knew all this. The creative problem for Archimedes was that nobody in those days knew how to calculate accurately the volume of an irregular solid without mutilating it. The story is that one day as he was taking a bath, Archimedes stepped into the full bathtub and it overflowed. This gave him the new idea that the volume of the crown could be determined by submerging it in water and measuring the volume of the water that the crown displaced. So energized was he upon his discovery that Archimedes is said to have raced naked into the streets of Syracuse, crying, "Eureka, eureka!"—"I found it, I found it!"

This was a discovery of a new context in which to study the equilibrium of bodies in a fluid as the buoyancy of the fluid supports them. The discovery was sudden, a discontinuous quantum leap, literally, in the thinking of Archimedes. There are no intermediate steps in discoveries such as that of Archimedes: A moment ago the thought was not conceivable; now it is.

What makes discoveries difficult is that jumping to a new context often requires a jump to an entirely new pattern of thinking. A single new thought does not cut it. You have to see a new gestalt of meaning and the new context that makes sense of that gestalt. So unconscious processing—processing without collapse and therefore without subject-object split awareness—is important.

Only unconscious processing—processing without collapsing every new thought in consciousness—can lead to a proliferation of possibilities for new contextual thinking. (A single thought would not make sense in isolation anyway.) Only when an entire pattern of new thought within a new supramental context has gathered in possibility is there an opportunity to recognize the pattern and take the quantum leap with your mind. Simultaneously, a state is collapsed in the brain that makes a map of the new mental context, a representation.

As an example of creativity, I will tell you how the previously mentioned creative quantum leaps occur in biological evolution. The idea is simple: Genetic variations (the mutations) are quantum processes (Elsasser 1981, 1982). They are not actual events but produce superpositions of possibilities, that is, possibility waves. If consciousness collapsed the possibility wave of each quantum mutation into actuality, the resulting variation would almost certainly be selected against, for individual mutations are seldom beneficial. So instead, consciousness waits until many possibilities accumulate and develop into a gestalt that consciousness recognizes.

Recognizes how? When the gestalt resonates with consciousness's purposive form-making ideas codified in the blueprints that Rupert Sheldrake (1981) calls morphogenetic fields. Out of the recognition comes choice and the collapse of an actuality that corresponds to an evolutionarily beneficial macroevolutionary trait. (See Goswami 1997, 2000, for further details.)

Once the brain representation of a thought is there, its subsequent activation always leads to the correlated thought. This is how we develop our learned repertoire of contexts that defines our ego.

The discovery of a new meaning in a new context is called fundamental creativity because this is what creativity is all about. However, we also engage in a poor facsimile of creativity in what is called situational creativity or invention. Situational creativity is the invention of new meaning in an old context or combination of old contexts. It is very useful when we have to deal with situational problem-solving. In biological evolution, when a species encounters environmental changes, how does it cope with it? Through adaptation—developing new traits using its reservoir of previously learned contexts, the gene pool. You can recognize Darwin's theory here. Darwin's is not a theory of evolution, but a theory of adaptation.

For situational creativity, consciousness is needed (otherwise who will see new meaning?), but not necessarily discontinuous quantum leaps.

Imagination plays a role in both kinds of creativity because in imagination, too, we try to move out of known contextual boundaries of thought. Imagination is new thought and making brain representations of it. It is not fundamental creativity because no discontinuity is involved, but it is a step toward creativity.

Meaning and Medicine

I read a story about a young man having an ongoing battle with his girlfriend on issues of sexism. One day while they were arguing, in the height of emotion she called him a pig. He left in a huff and while he was driving on a road that had meadows on both sides, another car passed him with a female driver in it. Just as she was passing him, inexplicably he heard her say the word "pig." He took it personally, became upset, and drove into a bush on the side of the road only to discover that there was a pig there, peacefully resting. Clearly, his fight with his girlfriend gave him a context for giving a meaning to the word "pig" that he could not help.

Nothing in the physical world has any inherent meaning; nothing that we see, hear, touch, taste, or smell has inherent meaning. An onlooker innocently utters the word "pig" on seeing a real pig, and somebody else interprets the word to mean "male-chauvinist pig," feels outraged, and that leads to an accident.

Apart from the physical, there is also the vital world of feeling. Do feelings—the movement of vital energy—come with inherent meaning? No. What gives meaning to our feeling? Mind does. Only mind enables us to process meaning. That is the job of the mind.

You are trying to soothe your boyfriend, who is irritable—his third chakra has an imbalance of vital energy. All of a sudden you notice he has managed to sweep you into irritability. What happened? Vital energy was exchanged between his navel chakra and yours, a nonlocal transfer, of course. Your mind automatically interpreted this resultant imbalance as irritability, so you, too, became irritable.

Does mind have a role in medicine? You bet. How we interact with and interpret the physical and vital world around us depends on the mental meaning we put to the stimuli we interact with. And the meaning can often have disastrous consequences on our health.

Item: I grew up in the snake-infested region of the world now called Bangladesh. Once in the dark as I was crossing the threshold of a room, I saw a cobra in a striking pose. Adrenaline must have rushed into my body and I jumped and ran. Later search showed there was most likely no cobra; I might have seen a rope, the search party said. But I could not sleep for days, and for years I was afraid of snakes; both consequences were tied to my mental, even physical, health.

Item: There is a famous story in the Bible. When Peter and his fellow Apostles were spreading the gospel and founding the Christian church, many householders were selling their houses and donating the proceeds for the good cause. Alas! Two of them, Ananius and his wife Sapphira, were keeping a bit of the take for themselves. When Peter found out, he admonished Ananius thus: "How is it that you have contrived this deed in your heart? You have not lied to men but to God." It is said that when Ananius heard those words, he died. His wife had the same fate when Peter confronted her.

Item: This is a case history published in a medical journal. John had a mastectomy to remove smelly abscesses in his right breast that had prevented him from doing any work for some time. Although the operation was successful and John did not have any known heart condition, he had a heart attack after the operation. He survived and came back home. But according to his wife, he was so disturbed that he still could not get back to work.

Things became worse when Halloween pranksters damaged his mailbox and an arbor that John had built. When John discovered these damages, according to his wife, he became depressed. He stared helplessly at his favorite arbor, said he did not feel well, and turned to go back to the house. Scarcely had he walked 20 yards when he collapsed. In another five minutes he was dead.

The cause of death was ventricular fibrillation, but what triggered it, according to the researchers who studied John's case, was John's feeling of helplessness and despair that had been growing for some time. It is the feeling that comes when we cannot find any meaning for our life. "To mean nothing is to die" is the subtitle of the chapter in which the physician Larry Dossey (1991) relates this story. Carl Jung said that meaninglessness is equivalent to illness. What meaning we put, even what meaning we don't put, to events in our lives affects our health.

Another way meaning is important for health and disease is how we look at our disease. It varies quite a bit from person to person, doesn't it? When I was young and under the constant pressure to perform in a career in physics, whenever I caught a cold or flu, I felt a little depressed about having to miss work. I am positive that this feeling actually interfered with my getting over the cold. I had a friend who was relaxed about a cold or flu and "enjoyed" the holiday it afforded him.

Dossey (1991) talks about his Vietnam experience in the same way. To some a disease was a nuisance and depressing, but to many of his comrades who wanted to escape Vietnam, a severe disease was welcome since it gave them the opportunity to return home.

Similarly, the pain researcher Henry Beecher found that during World War II, severely wounded soldiers often needed little help (such as morphine) for dealing with their pain. The reason is the same as for those in Vietnam. The soldiers were so relieved not to have to face the horrors of war; the wound meant to them freedom from the war.

How do the mind and meaning actually affect our physical body? Research is showing that they do so through the autonomic nervous system, the part of our nervous system that is involuntary. How do the mind and meaning affect our feelings, a process that I call mentalization? They do that through the autonomic nervous system and the recently discovered psychoneuroimmunological connections. This is discussed in the next chapter.

1 4

Mind as Slayer

When I was taking my master's exam in Calcutta, India, I didn't feel prepared for the exam at all; I became very anxious. The more I prepared, the harder I studied, and the more anxious I became. Then, the day before the exam, I developed a heart palpitation that wouldn't quit. Eventually, with the agreement of my family, I had to give up on completing the exam. Within a couple of hours of dropping out, the palpitation miraculously stopped.

This was my first experience with the effect mind has on the body, effects that are sometimes labeled "mind as slayer." There are many such effects. For example, many people report that when they are mentally down or depressed, they are more apt to catch a cold or something; it is reasonable to theorize that this happens because depression makes their immune systems vulnerable.

Not only does your mind have hazardous effects on your

health, but somebody else's mind can kill you if your belief system permits such a thing. It is documented that that's how voodoo works (Dossey 1991).

There are, fortunately, also mind-body effects called "mind as healer" (which are discussed in subsequent chapters), so there is recompense from the mind as well. But both these phenomena, mind-as-slayer and mind-as-healer, are suggestive of mind over body; they are new in Western medicine (Pelletier 1992) and are still distrusted by conventional Western medical practitioners.

What is the reason for the reluctance of the Western medical establishment in accepting and understanding mind-body disease or healing? It can be summarized in one word: dualism. Mind-body medicine, for a Western medical practitioner, conjures up images of a dualistic mind separate from the body (a discarnate soul, as it were) acting on the body. This is not palatable, because science elsewhere (physics and biology) is supposed to have eradicated dualism in favor of a monism based on matter—everything is matter (and its correlates, energy and force fields).

So the mind is looked upon as a part of the body—specifically the brain. In this view, the idea of mind causally acting upon body is circular—body acting upon body without a cause! Mind-body disease (or healing) is not permissible because of this circularity and breakdown of logic.

To state it for the umpteenth time, in a metaphysical foundation in which science is conducted *within consciousness,* separate mind and matter can work without dualism. Not so long ago, when asked, "What's mind?" the scientist would reply, "It doesn't matter." And when asked, "What's matter?" the scientist would retort, "Never mind," reminding you of the circularity of thinking about these things. But no more.

Mind and matter both are quantum possibilities for consciousness to choose from. "What's mind?" The new scientist says, "Mind consists of those possibilities of consciousness that, when collapsed, give you meaning and the experience of thinking." And to "What's matter?" the new scientist's reply is "Matter also consists of quantum possibilities of consciousness, those, when collapsed, that give you the physical sensations of seeing, touching, hearing, smelling, and tasting." Consciousness is clearly the mediator of mind and matter in their mutual interaction, and room is then made for mind-body disease as well as healing.

There are quite a few good books out now on the subject of mind-body medicine (see for example, Pelletier 1992; Goleman and Gurin 1993), but most of them suffer from a bad case of world-view jitteriness. There is a tendency in these books to pussyfoot around the idea of a causally potent consciousness and a mind separate and independent of the brain. I hope the psychophysical parallelism that I have outlined here cures this.

How the Mind Can Be a Slayer

Let's be more systematic and discuss in detail the solid evidence accumulating in favor of the idea that mind can cause disease. But first, what do we mean when we say that mind causes disease?

In quantum terms, mind helps consciousness process meaning. So disease, once again, can be due to faulty representations in the physical body (the allopathic model). In the mind-body medicine model, disease can also be due to faulty processing at the mental level, giving faulty meaning to physical events, mentalizing feeling—giving meaning to meaning-neutral feelings.

Much of the data of mind-body disease (psychosomatic illness) say it is stress related. First, we need to define some terms. A stressor is an outside agent, such as a death in the family, a math problem or an exam, a boring job, and so on. Stress is how a person reacts or responds to a stressor, which means what mental meaning he or she puts to the stressor and how he or she mentalizes the feeling associated with the reaction to the stressor.

Of course, in a given culture, meanings become somewhat fixed, in which case the stress associated with many common stressors produces a similar stress response in most people, so that one can talk about an average response. Important research by Richard Rahe (1975) measures such average stress in "life change units" (lcu)—the degree of life adjustments a stressor requires. For example, in Rahe's study, a minor illness has a stress level of 25 lcu, whereas the death of a spouse counts for 105 lcu.

Stress can cause heart attack. In this way even a seemingly simple stress such as an exam can kill a person. More people die of heart attacks on Monday than on any other day. This is called the "black Monday syndrome." Explanation? Monday is when we return from a relaxed weekend and we fantasize how awful the work we have to do is going to be, how boring, how difficult, and so forth.

Famous is the 1991 *New England Journal of Medicine* report showing a remarkable correlation between levels of mental stress and susceptibility to the common cold. In this research, psychologist Sheldon Cohen at Carnegie Mellon University injected volunteers with measured amounts of cold viruses or a harmless placebo. Among those who got the viruses (any of the five kinds offered), the chance of getting a cold was found to be directly proportional to the amount of stress in that person's life (according to their own estimate).

There is now at least preliminary evidence that stress can cause disease of the gastrointestinal system. An example is stomach ulcer. Never mind what some allopaths say, that ulcer is caused by a bacterium. Stress can cause severe diseases of the respiratory system (e.g., asthma) and the immune system (e.g., autoimmune diseases, in which the immune system attacks the body's own cells) and maybe cancer.

Cancer is the uninhibited growth of certain cells of the body. The immune system, when functioning normally, should be able to kill off such abnormally functioning cells. But the mind under stress leads to malfunction of the immune system, thus causing cancer. This may not be the whole story of cancer, but it is a plausible model.

Can the mind acting in conjunction with the brain affect the immune system? This question needs special attention.

Psychoneuroimmunology

Two great physicians of ancient Greece, Hippocrates and Galen, believed that thoughts and emotions moved to the various systems of the body and directly affected them via contact interaction. Today's avant garde research is establishing the truth of that; thus the preponderance of such terms as psychobiology and psychoneuroimmunology in modern medicine.

Imagine that following a certain stimulus, you are angry, very upset. Your mind is giving meaning to the stimulus you received that is producing your anger. Your brain is mapping your mind, but can the brain communicate its maps to the body, specifically the immune system? The answer is yes. The brain does that through recently discovered molecules called neuropeptides. The new wisdom gained—mind affects the brain affects the immune system—has become the subject of an entirely new field, psycho-neuroimmunology (abbreviated as PNI).

Before we delve into PNI, a little introduction to the immune system may be of help in order to understand why, before PNI, it was considered independent of the brain.

The organs of the immune system are also called lymphoid organs because they produce lymphocytes, the all-important white blood cells that are mediators of the immune response in the body. The initial production of the lymphocytes takes place in the bone marrow. One set of lymphocytes called T cells early in their development lives in the thymus gland (behind the sternum in the chest) and becomes the upholder of the "me-not me" distinction. Lymphocytes travel throughout the body, and of special importance are the small armies of these cells kept alert in the lymph nodes and the spleen. The immune system defends the body against intruders—viruses, bacteria, any foreign "not me" object. This seems independent of what the brain does.

The surprise came when a neurologist at the University of Rochester in New York discovered that all immune system organs have nerves all over them, so it is plausible the immune system communicates with the brain. Then neurophysiologist Robert Ader (1981) discovered that the immune system could be conditioned following the same procedure as mental conditioning.

You know the classic case of conditioning—Pavlov's fundamental research. Dogs are proffered food when a bell rings. After a while, the dog salivates whenever the bell rings, even without the food. In Ader's experiment, rats were used instead of dogs.

So let's consider the classic experiment of Ader that prompted him to coin the word "psychoneuroimmunology." Ader was working on a Pavlovian conditioning experiment of teaching rats an aversion to saccharine-flavored water. The standard practice was to correlate the rats' drinking of water with the taking of a drug (psychophosphamide) via injection, a drug that induces nausea and vomiting. Rats quickly learned to associate the sweet water with the nausea. After the conditioning, the rats would have nausea with just sweet water, and the drug was not needed any more. But there was a peculiar complication. The rats also seemed to have learned to die as a result of drinking sweet water.

Ader discovered that the drug induced a suppression of their immune systems. As a result of the conditioning, the rats had learned to simulate (upon drinking sweet water) not only the nauseating effect of the drug but also the immune suppression effect. It was the suppression of the immune system that made the rats prone to disease and death.

Experiments soon followed at the human level. One of the first such studies correlated the infection rate of sailors while onboard ship with their life events. The sailors who were the most unhappy as a result of their life events were also found to have the highest rate of onboard infection. Perceived negative meaning produced stress, which produced infection via the suppression of the immune system—a clear case of psychoneuroimmunology.

It is now recognized that stress (for example, produced by a spouse's death) can lead to reduced functioning of the immune system by reducing its arsenal of killer T cells. There is also little doubt that grief is a factor contributing to breast cancer in women.

Don't get too worried by all this talk about how stress negatively affects the immune system. Don't forget the previously mentioned Mother Teresa effect demonstrated by the study in which watching a film of Mother Teresa lovingly taking care of destitute, dying people increased the students' immune system function as evidenced by the increase of an immune enhancement marker (increase of salivary IgA). This is psychoneuroimmunology, too.

Molecules of Emotion

What mediates the interaction of the immune system with the brain? In the 1970s, Candace Pert (1997) and others discovered that the brain secretes certain molecules called neuropeptides, which help mediate analgesia, hormonal changes, and other responses to stress and resulting illnesses.

Among the neuropeptides, perhaps the most well known are the endorphins, which attach to specific receptor sites in the brain and the body (as in a lock and key mechanism). How the endorphins (or the lack of them) can alter our experience of pain (or pleasure) has been well covered in the popular press.

Take the case of hot pepper. Why is hot pepper pleasurable (actually what one experiences is pain mixed with pleasure) when from its molecular composition it should only give us pain? The answer is endorphins. Researchers prove this by using an endorphin blocker. If you take hot pepper along with an endorphin blocker, what you experience is unadulterated pain.

In 1979, it was discovered that certain components of the immune system, the T-cell lymphocytes, have receptors for an endorphin called methionine-enkephalin. This established conclusively that the neuropeptides, like endorphins, are the mediators between the brain and the immune system. Conversely, researchers found that the thymus gland secretes a substance called thymosin fraction 5, which stimulates adrenal hormones that have effects on the central nervous system. Brain endorphins connect to the immune system, and the immune system molecule thymosin connects to the brain.

The two-way-ness of the psychoneuroimmunological connection between the brain and the immune system was thus established. Similar two-way connections have now been established between the brain and the endocrine system as well.

Behavior or Beyond?

Journalist Bill Moyers did a series of TV shows on mind-body medicine (Moyers 1993) in which he showed the case of a Minnesota girl cured of lupus using Ader's PNI ideas. Lupus is an autoimmune disease that affects the connective tissues (tissue that connects, binds, and supports various structures of the body) and the blood. The problem with treating lupus is not that allopathic drugs that give relief from it are not known, but that all these drugs have dangerous side effects.

The girl was conditioned with the taste of cod-liver oil and the fragrance of roses along with the medicine, and the dosage of the medicine was gradually reduced as the conditioning took effect. In two years, the girl learned to manage with only half the dosage and eventually learned to produce the same effect as the medicine with just the taste and smell of the conditioning agents.

So what is going on? Conditioning to a stimulus is produced making use of our habit of experiencing stimuli only upon multiple reflections in the mirror of memory. Once conditioned, eliciting remembered behavior via reflection in the mirror of memory simulates the healing effect of a drug, even without administering the drug.

That sounds behavioral and brain-based, or does it? The questions are: Who learns initially? Who sees meaning? Who emotes? Who creates the memory? Who looks via the mirror of memory? The self is always lurking behind the behavioral language.

You hear somebody calling you an idiot, and you become angry. A trick of memory, certainly. But there is no effect if you don't know the meaning of the word "idiot." So your reaction depends on the conscious mental processing that you did originally with the meaning of this word.

Memories are representations of mental meaning. When we understand that consciousness and the mind are involved in making memories and looking through the reflections from the mirrors of past memory, we also recognize that consciousness and mind can undo the effect of the memories. This undoing of the body memory is the mechanism by which such techniques as massage therapy and Rolfing work (more on this subject in the next chapter).

As the machinery of the brain-body interactions is becoming clear, it is also becoming clear that this "machinery" is only an instrument for consciousness to use. Ultimately, it is consciousness that experiences moods and mood swings, emotions, stress, disease, and healing. To be sure, some of our conscious experiences are conditioned, but there is always scope for new choice, for creativity. The crucial role of creativity in healing is discussed in chapter 16.

Before we can understand healing, we must address the issue of how people process meaning, especially emotional meaning, why different people process meaning differently, and how this affects their health.

Mental *Gunas* and Mind-Created Physical *Doshas*

The usual approach of proponents of mind-body medicine is first to demonstrate with data that mind does cause disease, then demonstrate through a discussion of psychoneuroimmunology and such the mechanisms of how mind affects the body, and then go into the techniques of mind-body medicine. I think we can do better.

The straightforward approach does not explain why everybody does not contract mind-body disease, why the response to stress is not universal. There are data showing that people who are optimistic, committed to their work and have control over it, and look at stressors as challenges to be overcome do not suffer from the ill effects of stress (O'Regan and Hirshberg 1993). Among us, there are also Forrest Gumps, the Hollywood rendition of mentally slow people who breeze through life without feeling stress.

I have heard that Forrest Gump dies and is stopped at the pearly gates of Heaven by Saint Peter. "Not so fast, Forrest Gump. I am impressed that you had a full life without falling prey to the emotional distress of stress, but that is not enough. I must make sure that your mind works at least minimally. You have to answer three questions to prove your mental IQ."

"All right," says Forrest Gump.

"The first question," says Saint Peter. "How many seconds in a year?"

"Oh, that's easy," says Forrest Gump. "Twelve."

Saint Peter is puzzled. "How did you get that?"

"Count it. January the second, February the second . . ." Forrest Gump goes on counting.

Saint Peter stops him. "Okay, okay, I got it. I give you that one. Now the second question. How many days in the week begin with 't'?"

"Four," answers Forrest Gump.

"How do you get that?" asks Saint Peter, once more in puzzlement.

"Tuesday, Thursday, today, and tomorrow," Forrest Gump answers.

Saint Peter chuckles. "All right, all right. I will give you that one also. But this third one you must answer properly. What is God's name?"

"Andy," says Forrest Gump without hesitation.

"How do you get that?" asks an exasperated Saint Peter.

"Well, I learned it singing hymns at church. Andy [and He] talks with me, Andy [and He] walks with me . . ."

Saint Peter is amazed. "Well, I'll be darned; I will grant you that one also," he says, as he opens the gates.

I suspect that the lessons of individuality of vital body medicine—Ayurveda, traditional Chinese medicine, and homeopathy, all individually administered—are crucial here. There is individuality in our mental response to the stressors.

The question is this: How do we process our mind? Since mind is a quantum system, there are only three ways we can process it: fundamental creativity (the ability for quantum leaping from known contexts of mental meaning); situational creativity (the ability to create new meaning from a combination of known contexts); and conditioning (using known mental meaning). This gives us three qualities of the mind.

Recognizing these qualities of the mind was a profound accomplishment of Indian philosophy and psychology, in which the qualities are called *gunas* in Sanskrit. We have introduced this term before, in connection with how we process the vital body (see chapter 9). To avoid confusion, let's use the term mental *gunas* to denote the mental qualities. Easterners give names also to each of the individual qualities. The quality of fundamental creativity is called *sattva* in Sanskrit, situational creativity is called *rajas,* and conditioning *tamas.*

The importance of the concept of quantum qualities (mental *gunas*) of the mind—fundamental creativity *(sattva)*, situational creativity *(rajas)*, and conditioning *(tamas)*—can now be stated:

Their unbalanced use produces certain defects, *doshas* in Sanskrit, in the physical body, in the brain. We will call them mind-brain *doshas* (to avoid confusion with their vital body counterparts, defects or *doshas* arising in the physical body due to the unbalanced use of vital body qualities; see chapter 9).

It is not hard to see what these mind-brain *doshas* are. Overactive, unbalanced mental *sattva* creates the intellectual—one who discovers new contexts only for more thinking, not balanced living. In other words, an intellectual becomes detached from the body. James Joyce wrote an enigmatic line about a character in one of his novels: "Mr. Duffy lived a little distance from his body." This describes the intellectual perfectly.

I can't resist telling you a Nasruddin story in this connection. Mulla Nasruddin, a boatman in this story, was taking a pundit in his boat to a certain destination. As soon as they started their journey, the pundit started giving Nasruddin a sample of his knowledge, in this case grammar. But Nasruddin was bored and did not try to hide it. The pundit got irritated and retorted, "If you don't know grammar, half of your life is wasted." Nasruddin let the comment pass. After a while, the boat developed a problem and began to capsize. Nasruddin asked the pundit if he knew any swimming, to which the pundit replied no, adding that the idea of physical exercise bored him. Now it was Nasruddin's turn. Said he, "In that case, all of your life is wasted. The boat is sinking."

Overactive mental *rajas* give rise to hyperactivity at the physical brain level. Hyperactive people have a short attention span since situational creativity's demand for attention is considerably less than that of fundamental creativity; they also live a do-do-do lifestyle, being always focused on mental accomplishment.

Overactive mental inertia or *tamas* gives rise to mental slowness of the brain, a basic lethargy of the brain that keeps one from engaging in mental learning and processing.

Like the vital-physical *doshas,* these mind-brain *doshas* usually come in mixtures, giving us four more types: the hyperactive intellectual, the mentally slow intellectual (idiot savant), the mentally slow hyperactive, and the mixture of all three.

Although these mind-brain *doshas* reside in the brain, they govern our attitude toward all emotions. Of people of the three *doshas*, only the mentally "lazy" avoids the mind-brain duo and lives in the body, and not only in the three lower chakras, but also in the heart. Persons of the other two *doshas* mentalize their feelings. People of predominant intellectuality will suppress emotions and are prone to suffer from chronic depression as a result. People of dominant rajasic *dosha*—hyperactivity—are of the expressive kind; they are easily irritable and will be prone to quick anger and hostility in their reaction to stress. Hyperactivity may also be associated with anxiety.

In India, you get to wait a lot at airports because planes are seldom on time. To pass time, I sometimes watch people, and it is interesting how very quickly I find verification for the threefold classification of the mind-brain *doshas*. Some of the people will look stoic, but if you give them an opportunity, they will immediately start grumbling. These are the intellectuals. Then there are those whose anxiety shows; they are impatient and restless, very prone to bursting out in a bout of anger. These are the hyperactives. But some people are content with the situation and appear to be stable. It is not necessary to assume they have arrived at the much-coveted state of mental equanimity. No, these people are just mentally slow to process things.

Movies such as *Forrest Gump* seem to portray the idea that only the simpleton can be happy, can be nice to others. There may be some truth to that in driven cultures like that of the United States, because most people in this country suffer from overactive *sattva* or *rajas* (or both) and thus from the *doshas* of intellectualism and/or hyperactivity, whereas only a few get to "enjoy" the niceties of *tamas*, mental slowness, at the brain level.

Finally, when considering correcting mind-brain *doshas* we should remember the concept of *prakriti* (see chapter 9). We all have (due to reincarnational and early developmental proclivities) a natural homeostasis of all three *doshas*, although one or two *doshas* dominate. This is *prakriti*. It is the deviation from *prakriti* that gives us health problems, and it is that which needs to be corrected.

The general rule of thumb is this. Excess intellectualism has the tendency to suppress emotions. Excess hyperactivity leads to the tendency to express emotions. This is further discussed in the next section.

Response to Emotion

How do we respond to emotions? In the West, especially in America, there is strong cultural conditioning against expressing emotions. Expressing emotions is considered a sign of weakness and hence, almost universally, Western men learn to suppress emotions. For women, on the other hand, the cultural conditioning against expressing emotions is not as deep.

Nevertheless, not all Western men suppress their emotions. For example, if one has an exaggerated notion of self-importance, one indulges in expressing emotions, not needing the usual social constraint of defending one's persona. You can see such people everywhere. Under emotional stress, these people have well-recognized responses of short temper or irritability. We can see the connection with mind-brain *doshas* here. The intellectuals almost universally manage to suppress their emotions. But not everyone with dominant hyperactivity as their mind-brain *dosha* suppresses emotions. This is especially true when excess hyperactivity, out of balance with one's *prakriti*, develops. Thus the *dosha* of hyperactivity, when in excess, can easily result in expression when confronted with emotional stress.

Here is something else. If one is fortunate, there is always another person to allow the ventilation of emotions; this other person can help dissipate the negative impact of the emotional expressions. In traditional societies, this used to be the rule, so the health impact of emotive expression was relatively minor. But now it is all changing.

What does unsupported expression of emotions under mental and emotional stress do to us? This is now fairly well understood (Goleman and Gurin 1993). Response to stress is a function of the autonomic nervous system, and this system has two components, sympathetic and parasympathetic. As its name implies, the sympathetic nervous system sympathizes with us and brings to bear the change in physiology that we need to "survive" the stimulus responsible for the stress. The parasympathetic system controls the "relaxation response" designed to bring the body back to equilibrium.

So what does prolonged exposure to stress stimuli do to us if we allow emotional expression in response to an excess of the mind-brain *dosha* of hyperactivity? In general, expression produces an imbalance in the activities of the sympathetic and parasympathetic nervous system so that the end result is that the system is left in a permanent state of sympathetic arousal.

What happens then? Chronic irritability and nervous tension can lead to sleeplessness. This is only the beginning. Chronic irritability, which comes from hurriedness, combined with competitiveness gives rise to hostility. Eventually, what was previously mental hyperactivity expressed through conditioned programs of the brain becomes manifest in the physical organs, which all begin to function at a hyperactive level, producing disease of these organs. Often, the disease settles in one organ only.

In this way, chronic arousal due to the expressiveness of the emotional response has been associated especially with heart disease and hypertension. But heart disease is not the only result of this expression. If the expression occurs through the gastrointestinal digestive system, the result is ulcer. If the expression is through the elimination systems of the body, the diseases are irritable bowel syndrome or bladder disorders. If the expression is through the immune system, causing excess immune reaction to antigens, the result is allergy. If the expression is through the respiratory system, the disease is asthma. And so on it goes.

Why does the expression settle in one organ as opposed to another? This is the million-dollar question. I am convinced that this has to do with the response of the vital body where feelings originate in the chakras.

Recall that different types of emotion are felt at different chakras. For example, irritability and anger are felt at the navel chakra. This happens when we are not getting what we want, when our ego is being affronted. The mental stuff, the processing of meaning as according to the mind-brain *dosha*, amplifies the vital feeling at that chakra. In this way, chronic irritability expresses itself in the organs of the navel chakra, most often as peptic ulcer.

But when irritability gives way to hostility, an advanced response of people with excess mind-brain hyperactivity, where is the vital energy felt? Hostility is looking at the world as enemy, as *not me*. This happens when vital energy in the heart chakra is depleted and goes into the navel chakra. Thus hostile reaction inevitably leads to disease of organs in the heart chakra. If the hostile reaction is directed at people, the organ affected is the heart. If the hostile reaction is directed toward the environment, the organ affected is the immune system.

Instead of hostility, advanced stages of irritability and competitiveness can also give rise to frustration, which is a throat chakra feeling (arising when the throat chakra is depleted of vital energy). When mind gets into the action, the feeling of frustration is amplified. Repeated amplification of frustration expresses itself as a throat chakra disease, e.g., asthma.

If the emotion expressed is fear or insecurity, the chakra involved is the root chakra. When amplified by the mind, this may lead to problems of the root chakra organs, such as diarrhea and irritable bowel syndrome.

When the feeling involves the sex chakra, as in the feeling of unfulfilled lust, mental amplification gives rise to diseases of the second chakra (sex chakra). The enlargement of the bladder, responsible for urinary problems in many males age 60 and older, is a disease of this kind.

Suppression of Emotions

What does suppression of the emotional response of the mind do to us? As Freud and psychoanalysts correctly theorized, mental suppression or repression may be represented (psychoanalysts call this "conversion") as certain brain states that cause (false) physical symptoms even though there are no physiological changes. This has been called a defense mechanism, because the mind and the brain "defend" the organism from embarrassment from a cultural point of view. This, of course, is what was originally recognized as the only psychosomatic disease—disease without any physiological basis. (Unfortunately, many people still think of all mind-body disease as without physiological effects.)

We can try modeling "real" psychosomatic disease in the same way. For example, the psychiatrist John Sarno writes:

> My patients have shown that the underlying psychology is the same for conversion and mind-body disorders. It is as though the brain has decided that the conversion symptoms were no longer convincing as disease, so it began to produce processes in which there were obvious physiologic reactions. This was done by involving the autonomic and immune systems in the production of symptoms (Sarno 1998, p. 46).

However, in such a model, a question is left unanswered and mysterious: How does the brain, being a machine, decide to produce a psychosomatic reaction and how does it choose its location?

Seeing the role of consciousness in decision-making and the role of the vital body in the choice of the location of the psychosomatic reaction in emotion solves this mystery. Every emotion has a vital-body counterpart, a feeling associated with it. The feeling is connected with vital-physical body movements at a certain chakra. The physical representations of the vital movements that we feel involve the corresponding organ(s) and also the muscles in which the organs are embedded. When mind suppresses the emotion through the intermediary of the brain and its connection to the physical organs through nerves and neuropeptides, the vital body movements in the corresponding chakra are suppressed along with the programs that run the functions of the physical representations, the organs. This is what is responsible for the somatic effect, the experience of illness at a specific organ site because of an actual change in physiology there.

Psychiatrist Wilhelm Reich was a specialist in the problem of suppression due to intellectualism for which he described the symptoms thus:

> . . . loud, obtrusive laughter; exaggeratedly firm hand-shake; unvarying, dull friendliness; conceited display of acquired knowledge; frequent repetition of empty aston-ishment, surprise, or delight, etc.; rigid adherence to def-inite views, plans, goals; obtrusive modesty in demeanor; grand gestures in speaking; childish wooing of people's favor; boastfulness in sexual matters; display of sexual charm; promiscuous flirtation; pseudo-exuberant fellow-ship . . . (quoted in Grossinger 2000, p. 433).

How does a therapist deal with this kind of behavior in a patient? The Reichian therapist's answer is by confronting. Here is an example:

> For instance, the patient may acknowledge that, of course, he has shown contempt for people in the past, that aloofness is one of his vices. The therapist may answer, "No, right now. You are being contemptuous right now. Even as you pretend to participate in this session, your expression says that you are merely indulging me, showing contempt for me." If the outraged and embarrassed patient denies this, the therapist can respond, "Feel your mouth. Your lips are frozen into a leer. You have no sensation of them any-more." Upon checking the patient is astonished to realize this is true. Automatically, he grins foolishly. One therapeu-tic method might then be to massage the lips in order to restore feeling (Grossinger 2000, p. 434).

This emotional repression by the mind, by mentalization—giv-ing meaning to a feeling and, in this case, making it into some-thing to be avoided—when chronic, becomes repression of organ functions at the chakra corresponding to that feeling. Crucial in this process is the connection of the brain to these various organs via the nervous system and neuropeptides (as in psychoneuroim-munology).

In particular, if the immune system function is suppressed, we may get various autoimmune diseases. Cancer may be the result of reduced immune system activity. Suppression at the brow chakra may be responsible for tension headaches and migraines. And suppression at the crown chakra leads to depression at the psychological level (see also Page 1992), which is a contributing factor in chronic fatigue syndrome.

However, repression most commonly is memorized in the muscles. This is because when we are defensive, we tend to tense our muscles. As we repress the mental-emotional experience, we also repress the muscle tension and never fully relax the muscles. In this way, repression of the mind translates as the repression of muscular activity. The muscles retain a "body memory," so to speak, of the emotional trauma suppressed. I think it is fair to say that a muscle is holding a memory when the muscle is fixated in a certain position and cannot relax that position.

The physicist Fred Alan Wolf (1986) clarified the mechanism of how the muscles retain memory. Each muscle, which is an array of long cells up to one foot in length, contains many cell nuclei and many small fibers called myofibrils. Myofibrils are made up of repeating units of sarcomeres in a lengthwise arrangement along the long cylinder axis of the muscle. Muscle bioenergetics depend on the free flow of calcium ions. When a muscle is tensed (as when an emotional trauma is being defended against and suppressed), the muscle sarcomeres are flooded with these ions of calcium. Even after the traumatic incident is over, some excess calcium may remain in the sarcomere. This continues to maintain the tension in the muscle, becoming a memory of the suppressed trauma.

What does repeated suppression of an emotional response mean in terms of muscular tension memory? Quantum mechanically speaking, in subsequent experiences of that stimulus, just as the mind is not allowed to collapse certain mental states of awareness of the emotional response, the particular muscle memory is never "collapsed." So this particular muscle is not reactivated by subsequent emotional experiences if the mental defense mechanism is always aroused.

It is likely that suppressed emotions all over the body give rise to serious diseases such as fibromyalgia, a state of widespread muscle pain. A related disease is called chronic fatigue syndrome, in which the main physical body symptom is total fatigue. If feelings are suppressed in all the body chakras, practically all the corresponding vital body movements will be suppressed. This may manifest as a general lack of vitality, explaining chronic fatigue. If the feeling suppression involves more of the structural parts of the body in which the organs are embedded, but not the organs themselves, the lack of vital energy may be felt as pain all over the body—fibromyalgia.

A comment about pain. A past issue of *Newsweek* had a banner headline on the cover that promised breakthroughs in our understanding of pain. A look at the article inside, however, was disappointing. The article fell short of reporting any real breakthrough. Fibromyalgia, the article said, is real, because a new imaging technique (MRI) confirms that when a patient of fibromyalgia is crying out in pain, certain brain areas become active. Good. But then the article has nothing much more to report except that there may be a genetic connection to fibromyalgia.

Pain is interesting because being a feeling it must have a vital energy connection, yet the role of the nerves is also undeniable, since by numbing them (local anesthesia), we can numb pain also. So pain is a mentalized feeling, a feeling connected with the suppression of vital energy at any structural part of the body and interpreted by the mind as pain, because it is undesirable. This is a very persistent mentalization, obviously millions of years old, and has much survival value.

The Disease-Prone Personality

Is there such a thing as a personality type that develops a specific mind-body disease? For example, coronary heart disease is connected with the type A personality, people who react quickly with anger and hostility, especially hostility, to a stress-producing situation. Is such a connection valid for other kinds of disease in which the mind may be involved?

At one point, there was a considerable amount of literature connecting cancer to the type B personality, associated with the inability to express emotions and nonassertiveness, even hopelessness. But such a connection has not been clinically demonstrated without controversy.

Thus there is no typical cancer-prone personality. This is understandable. All people of type B personality do not get cancer because, for some, the type B personality is close to their *prakriti,* their nature; even type A people can get their immune systems out of order and hence get cancer if they direct their hostility inappropriately.

Conversely, it is also true that not all cancers originate at the mental level, as I have been saying all along. Some cancers originate at the vital level; they are due to vital energy imbalances at the heart chakra. Some cancers have a genetic origin, and there are any number of combinations of physical, vital, and mental imbalances that can contribute to cancer.

There is clinical support (Freedman and Booth-Kewley 1987) for the idea that there is such a thing as a disease-prone personality. Freedman and Booth-Kewley were motivated to study the specific connection of personality types with asthma, coronary heart disease, ulcers, headaches, and arthritis. They found little evidence of any specific connection of any of these diseases to a personality type. Instead, their data showed the existence of a disease-prone personality with characteristics such as depression, anger/hostility, and anxiety.

I think this finding is entirely consistent with the idea of mind-brain *doshas.* As mentioned earlier, some people have a mixture of mind-brain *doshas* that manifests as a personality with more than one predominant disposition toward emotion, both suppression of emotion (depression) and expression (irritability, hostility, and so on). All we can say about such people is that they are disease prone.

Here is a question: If I have a disease-prone personality, doesn't it seem that I am responsible for my disease? Should I then feel guilty?

Many New Age teachers will squarely put the blame on your shoulders for your ailments (Why are you hiding behind your heart disease?), but in truth, do we really know that your disease has been produced at the mind level and not at the vital or the physical level? (And even if it is at the mind level, it is conditioned mind that is responsible. You are a little helpless.) The fact is, we don't usually know; we cannot know without the power of deep intuition.

At the same time, what prevents me from taking responsibility for healing myself if I do want to be healed? When I take such responsibility, only then can I truly engage with the advanced techniques of mind-body healing (see chapters 15–17).

Unnecessary Mentalization of Feelings Can Be Harmful to Our Health

A feeling is a feeling is a feeling. It is not inherently good or bad. The values we give to feelings, our likes and dislikes, are mind-created through the mind's "job" of giving meaning to everything it is able to process. This is one way we mentalize our value-neutral feelings.

Anthropologists have found that some Eskimo natives they encountered did not have a word for anger. This must mean that anger, as an emotive expression, was not a part of the social world of these Eskimos. Of course, this changed after their interaction with American anthropologists began: They had to coin a word for anger to describe the irritability and frustrations they saw in the behavior of the anthropologists.

Consider a feeling like fear. If a tiger comes into my periphery, fear, the draining of vital energy out of my root and navel chakras, gives me the physical adrenaline rush that helps my "flight" or (rarely, in this case) "fight" response to a tiger in my den. It is a necessary feeling, necessary for the survival of our species, and, no doubt, Darwinian evolution has helped to make it an instinct. But what if I fantasize a tiger in my living room and become afraid as a result of my fantasy? I may get a shiver through my body and butterflies in my stomach because of my fantasy fear, and even an adrenaline rush, but it is a case of mind over vital body, an unnecessary mentalization of an otherwise useful natural feeling (see Dantes 1995).

Something like the September 11 destruction of the twin towers of the World Trade Center in New York happens, and fear, indeed, is an immediate natural reaction for New Yorkers. It protects them from the immediate danger that the events might have meant—more terrorist assaults. But then the media react, the politicians react, the event is played on TV over and over, and what happens to the national psyche? Many children in New York, many people around the whole country, could not sleep at night for months after because they continued to suffer from fear because of September 11. This is an unnecessary mentalization of fear on a large scale, driven by media and politics.

The Bottom Line

The bottom line is this: Mind gives meaning to both the physical and the vital. How the mind processes meaning depends on the three qualities we bring (maybe even from past lives): *sattva* (fundamental creativity), *rajas* (situational creativity), and *tamas* (conditioning). These create three mind-brain *doshas:* intellectualism, hyperactivity, and mental slowness in the brain processing.

The first two *doshas,* when aggravated beyond *prakriti,* the natural homeostatic level, produce a tendency for the suppression and expression, respectively, of emotions. Both of these tendencies can give rise to disease at the physical level. In addition, unnecessary mentalization of feelings, giving meaning and value to feelings generated by mental fantasies, may produce disease in people of all three mind-brain *doshas.*

How Should We Deal with Meaning?

Knowing that meaning contributes to our illnesses and diseases, if you fall sick, you may be tempted to contemplate whether the sickness could have been caused by you, whether you are to be blamed. Alas, this will only further aggravate your situation.

If meaning is something inherent in how the mind processes things, if we are helpless, giving disease-causing meaning to our experiences in the world, including the diseases we suffer through, what is our best strategy in dealing with the mind? Some people say that thinking of disease objectively is the best from this point of view alone. But as Dossey (2001) correctly points out, denying meaning is also assigning meaning, a negative meaning.

Denying the meaning of our illnesses is like atheists denying the existence of God. Now if we could be truly agnostic!

So what is a good strategy? As Epictetus said, "Things in themselves are always neutral, it is our perception which makes them appear positive or negative." If we put a negative mental meaning to an event, it causes an incongruence with our normal state of happiness. Instead, suppose we interpret everything so that congruence is maintained?

The East Indian mystic Swami Sivananda gave a wonderful overall strategy for dealing with the meaning-giver mind, which I will share with you.

A king had a companion/minister whom he liked very much except for one thing that irritated him to no end. The minister had the habit of saying, "Everything that happens is for the good," to whatever happened around him, good or bad. So one day, the king cut his thumb while playing with a knife, and the minister, who was there, promptly said, "Everything that happens is for the good." This comment made the king very angry and he threw the minister into jail. To console himself, he went hunting in the forest alone.

He must have gone quite a distance and beyond his kingdom, because he came across a tribe that took him captive. Unfortunately for the king, this was a tribe who offered humans as sacrifices to their deity. So the king was taken to a priest to be offered as a sacrifice. The priest, while bathing the king, discovered his cut thumb and, since a defective person cannot be offered to the deity, rejected the king, who was then released.

While returning to his palace, the king had second thoughts and he realized that the minister's saying was correct. Indeed, the cut thumb had saved his life. So as soon as he was back, he released his minister and said to him, "You were right about me; everything that happened to me was for my good after all. But I threw you in the dungeon for what you said; so it did not seem to do you any good. How do you explain that?"

To this the minister said, "Oh, great king. Your throwing me into jail saved also my life. Otherwise I'd have accompanied you hunting, been taken captive, and since I do not have any blemishes, would have been offered as the sacrifice."

In Brief

This chapter has been about how the mind causes disease. Take the following summary information with you for further deliberation and application:

- How your mind reacts to stress-producing agents (the stressors) determines whether their effects are going to be adverse or otherwise.

- Your mind affects your brain, with which it is correlated via consciousness. Your brain is connected to your immune system via the movement of neuropeptides. In this way, your mind, via the brain, can affect your immune system.

- It is you who assign meaning to stimuli and it is you who choose to bury an unpleasant stimulus in your unconscious. Admittedly, it is the conditioned you, but you have the choice to come out of your conditioned cocoon if you so intend.

- How you process the meaning of your emotions depends on your mental tendency or quality, whether you have a predominant propensity of fundamental creativity (*sattva* in Sanskrit), a predominance of situational creativity (*rajas* in Sanskrit), or whether conditioning (*tamas* in Sanskrit) dominates you. These qualities, *sattva, rajas,* and *tamas,* are your mental *gunas* with which you process your quantum mind.

- If your dominant quality for mental processing is fundamental creativity, *sattva,* you develop the dominant mind-brain *dosha* or defect of intellectualism. If your dominant mental quality is *rajas,* you develop the dominant mind-brain *dosha* of hyperactivity. And with *tamas* as your dominant mental quality, you develop the dominant mind-brain *dosha* of mental sloth. If you have a mixture of two mental *gunas,* you develop a mixture of two dominant mind-brain *doshas.*

• What is (are) your dominant mental *guna(s)?* What is (are) your dominant brain *dosha(s)?*

• Every healthy person has a natural base level (homeostatic level) of the three brain *doshas,* called *prakriti* in Sanskrit. Can you figure out what your natural base level *prakriti* of brain *doshas* is?

• Excess mind-brain *dosha* of intellectualism is associated with suppression of emotions and the type B personality, which is sometimes linked to cancer, although making this link is controversial.

• Excess mind-brain *dosha* of hyperactivity is associated with the type A personality, and all studies agree that excess hyperactivity can lead to heart disease.

• Excess of both mind-brain *doshas* of hyperactivity and intellectualism may be the cause of a disease-prone personality.

• It is best to avoid the mentalization of feelings—putting meaning to feelings. In addition, health requires a minimum of fantasizing that leads to negative emotions—feelings that are detrimental to our mind and physiology.

• The best strategy for a healthy mind-body relationship is to see the positive side of everything. Develop this capacity. It will help you to surrender the ego and make room for the quantum self.

---
1 5

The Quantum Explanation of the
Techniques of Mind-Body Medicine

Knowing how mind-body diseases are produced, the basic ideas of preventing them should be fairly obvious. This should help us to understand the efficacy of the techniques that are in vogue now.

The techniques I discuss here are at least decades old, and some of them are very old, practiced for millennia. This has caused a problem. Because of overfamiliarity, and because of their demonstrated empirical validity, some of us (the believers) take the theoretical validity of these techniques for granted. Of course, the opposition camp doesn't subscribe to the belief in their validity; in truth, their validity, in the theoretical sense, has never been fully demonstrated. I will show in this chapter that when you add quantum and primacy-of-consciousness thinking to the scene, the empirical validity of these techniques is augmented by theoretical validity also.

At the most superficial level, we treat the symptoms of the mind that are the closest correlates of the disease. This is similar to allopathic medicine, in which one treats the symptoms of the body instead of those of the mind. You have a problem with hostility, which is "causing" (or may cause in the future) your heart problems, so modify your behavior. Your physical ailment is being "caused" by your negativity; try to change your negativity using the power of positive thinking.

Of course, if your belief system is basically materialist, then you don't believe in the causal efficacy of consciousness, and mind to you is just part of the brain. Then behavior is what matters and behavior modification is your only tool for mind-body healing.

Does behavior modification work? Can hostile behavior, for example, be changed through "reprogramming" the mind? Can good thoughts keep us away from violent thoughts? Can you reason yourself out of hostility? People have been trying to change violent behavior this way for millennia with no discernible result. Violent behavior does not change. Face it: At best, behavior modification techniques provide people with coping mechanisms. They help us to cope with situations of minor upsets. They are better than doing nothing; they buy you time, but that's about it.

You have to go beyond behavior modification and positive thinking. Suppose you validate your mind as separate from your brain. Suppose you open your mind just a little to permit the causal efficacy of consciousness—this is downward causation. Then you can let your consciousness and mind explore new states that have not been mapped in your brain yet, new states that contain states of health that you allow yourself to imagine or visualize (if you are good at visualization). Two notable therapies that can help you are hypnosis and biofeedback.

Hypnosis and Biofeedback

What is hypnosis? It is the power of our consciousness to pay attention to states of the mind so as to shift our identity, in this way avoiding the specific ordinary ego demands of the physical world upon us. Since the ego individuality is not real, but an identity that consciousness puts on, consciousness naturally has the ability to shift this identity. So the power of hypnotism is easily accommodated in a science within consciousness.

The most commonly acknowledged form of hypnosis is when somebody else (a hypnotist) helps us get to these "other" mental states in preference to our ordinary state of experience, including our ordinary identity. But self-hypnosis, getting to these hypnotic states by ourselves, is actually much more common. In fact, spiritual teachers have told us for millennia how we avoid reality through self-hypnosis created by our conditioning, the habit of processing everything through reflection in the mirror of our memories.

Say somebody or some stimulus elicits a hostile reaction in you. Instead of thinking good thoughts, if you use self-hypnosis to take you to a state of intense relaxation, is that a better strategy for changing your hostile behavior? The answer is a definite yes. In this way hypnosis, both induced with the help of others and in the self-induced form, can be used to control negative behavior.

Can hypnosis be used as a therapy for mind-body disease—for that matter, for any disease? Again, the answer is yes. Research shows that the regular practice of self-hypnosis, especially when used as a complementary technique, helps one to manage pain, manage blood pressure, manage a relative stability of functions controlled by the autonomic nervous system, stabilize blood sugar in diabetics, even reduce the acuteness of asthma attacks (Goleman and Gurin 1993). This is a very good performance record for self-hypnosis.

On the negative side, if one feels reluctant to acknowledge one's power of imagination, then hypnosis is less likely to work.

Fortunately, the other technique, biofeedback, can work for virtually everybody, but one has to be motivated. The idea is simple. There are many neurophysiological or biological functions that can be monitored and amplified by a machine, and the data can be fed back to the person using any one of the five senses. By studying the feedback and seeing how your conscious states of experience affect the neurophysiology (what is being fed back), gradually you can learn to exercise voluntary control over some of the functions of the autonomic nervous system. Eventually, you learn to achieve deep states of relaxation from which it is easier to control and regulate the autonomic functions, initially through the feedback machines, but gradually even without the use of the machines.

The basic idea, of course, is that consciousness acting on mental states can affect physiological states, and vice versa. The change of physiology will bring about a change in the conscious mental state. This idea may be anathema to the strict materialist, but is part and parcel of science within consciousness where psyche and soma are both recognized only as correlated possibilities of consciousness. Changing one naturally changes the other. Like hypnosis, biofeedback has also been used in therapy to help with chronic disease conditions or problems, such as headaches.

Releasing Memory

Once we permit the primacy of consciousness, it is easy to think about another kind of mind-body healing. In psychoanalytic theory, it is assumed that people often repress the memory of a childhood trauma in their unconscious; later, unhealthy behavior arises from the processing in the unconscious. But since the subject is not aware of where her behavior is coming from, she cannot do anything about it.

The job of psychoanalysis in mind-body healing is to make such unconscious memories conscious through therapy. More recent forms of psychoanalysis—for example, psychodynamics— are designed specifically to explore current emotional reactions (such as hostility) in terms of past memory.

Does psychoanalysis work? Behaviorists sometimes pooh-pooh psychoanalysis because it implicitly acknowledges the causal efficacy of consciousness and also because in the behavioral/cognitive worldview, it is impossible to distinguish between the conscious and the unconscious. Quantum physics allows us to understand the distinction between what psychologists call the unconscious and consciousness (see chapter 6).

The unconscious is consciousness without awareness, when no quantum collapse takes place. Consciousness refuses to collapse traumatic memory because of the pain involved. Therapy can help to relax the fear of pain and so when the traumatic memory surfaces, consciousness is able to recognize, remember, and become aware of it. The healing power of such awareness can be enormous (see Sarno 1998).

I mentioned in the previous chapter repression of traumatic memory in the body in the form of uncollapsed states of skeletal muscle excitations. In the East Indian technique of hatha yoga, the yoga postures are designed for releasing uncollapsed muscle tensions by becoming aware of them, thus relieving pain. Recent techniques such as Rolfing are designed to work in the same way.

Chakra Psychology

Next is chakra psychology. Chakra psychology uses psychotherapy to remove a blockage or imbalance of vital energy at individual chakras.

Some psychologists go too far in suggesting that all diseases have one ultimate cause; we cause them through confusion of our intentions. But we can't; in our ego we do not have the power of intention or downward causation. Instead what happens is that our conditioned thoughts inappropriately amplify the movement of vital energy in and out of the chakras, adding to an already existing disharmony at the chakras. Chakra psychology attempts to harmonize this disharmony. Below I give a chakra-by-chakra description of how chakra psychology works for healing (for further details, see Page 1992).

If the disease involves the root chakra (elimination systems), then the problem is insecurity, not enough healthy grounding. Remember, in our culture we try to ground ourselves through watching sex and violence on TV, but this is unhealthy grounding. What is healthy grounding? Simple tasks like gardening or walking barefoot on the ground can help us ground, but this is working with vital energy at the vital level. To work on the vital energy imbalance through the mind, we can use imagination and visualization. For example, close your eyes and imagine that there are roots from your root chakra that extend all the way to the center of the Earth.

For the second chakra, we can use our sexual relationships to rebalance a vital energy imbalance. Have you been ignoring your feminine side (if you are male) or masculine side (if you are female), the parts Carl Jung calls your anima and animus? To integrate the male and the female within you (which balances sexuality) during sexual union, visualize yourself as both male and female. This practice can be done even without sex.

For navel chakra work, how do you deal with mental amplification of anger and irritation that then becomes chronic? If you look at it, you will see how your mental hurriedness and impatience are major factors. So the psychological work here is to slow down (discussed later).

Hostility is a contraction of the energy at the heart, so the basic objective of psychotherapy at this chakra is to expand the heart. As Swami Sivananda, a spiritual master in India who lived in the last century, used to advise, "Be good, do good." This expands the heart. Also loving yourself frees you to love others and banishes hostility.

Traditionally, the heart chakra is depicted symbolically as the opening of a lotus flower. The lotus grows in the mud and is thus a good symbol for the transformation of the negative that is a potential for heart chakra work.

Often other people's vital energy (nonlocally) hits us at the heart chakra, especially when we are sympathetic. If that is the case, we must try to be objective and not identify with the other's troubles (empathy). Additionally, we may visualize putting a reflector around our body to reflect back all the received vital energy whenever we interact with another of negative energy.

Visualization is of great help in dealing with the suppression of the immune system. You can, for example, visualize energetic killer T cells fighting with intruders in your body and winning the battle. Sometimes this is demonstrated by visualizing your headache and making it smaller and smaller. Indeed, for some people, the headache *is* literally visualized away. Guided imagery is routinely used for the relief of chronic pain and for accelerating healing and easing discomfort from injuries (Achterberg 1985). Visualization is useful for working with chakras, in general.

For the throat chakra imbalance of vital energy, to deal with frustration of expression, the psychological task is to find avenues for creativity. If avenues in the public arena are too difficult to open (not everybody is born with talent), then engage in private smaller areas of creativity. For example, be creative in gardening, cooking, relating, singing in a small group, going out dancing, writing in a diary, understanding scientific ideas.

In dealing with the blockage of vital energy in the brow chakra having to do with suppression, the question to ask is: What is this block keeping me from doing? The answer, of course, is that it is keeping you from the full expression of your possibilities by denying the facility of focusing. (When we have headaches, we cannot focus.) Mentally, you are taking yourself, your learned repertoire of expertise, too seriously. Lighten up. There are many more possibilities to explore; learn to play with them.

I am happy to report that our new science conferences are often injecting humor and laughter as a regular event. At the European transpersonal psychology conference in Assisi, Italy, in 2000, we participated in laughing meditation for half an hour every day. And who hasn't heard Swami Beyondananda making fun of New Age researchers at the Science and Consciousness conferences in Albuquerque?

To work with blocked energy at the crown chakra, which leads to depression, one of the best psychotherapeutic tools is meditation (see the following section) on peace (Goswami 2003).

Crown chakra work is very important. Remember that liberating song in *My Fair Lady:* "The rain in Spain stays mainly in the plain." To the same tune, appreciate that "the pain of strain is mainly in the brain." This is because of brain's direct correlation with the mind. So we must do vital-energy practices to augment the meditation on peace, practices such as hatha yoga, *pranayama*, tai chi, and others (see chapter 11).

A common method for healing all the chakras is to visualize regularly healthy vital energy at each chakra. Recall the chakra meditation with a partner in chapter 11. This meditation, which also involves a lot of visualization, can be used to heal the chakras.

Meditation

At the next level of sophistication, if we are not a staunch materialist, we can look into the cause of the behavior that is causing the health problem. At that level, we are ready to deal with the cause of our behavior, the mind-brain *doshas:* excessive intellectualism and excessive hyperactivity.

How do we deal with excessive intellectualism? Intellectualism keeps us away from the body, away from experiencing the emotions. Instead emotions become something of a nuisance, something to be ashamed of, something to suppress at all cost. The remedy is, of course, to indulge in the body. Exercise is good, massage is good, hugging people is good.

Hugging works. Many years ago, I was an intellectual. When I was doing intense spiritual work in the 1980s, the mind-brain *dosha* of intellectualism, although not yet a health problem, became a problem for me against spiritual opening. I remember going to a workshop, and the workshop leader (the physician Richard Moss) prescribed "juicy physicality" for me, to be administered as hugs from my fellow workshoppers. It worked.

A complementary technique is meditation with the purpose of becoming aware of feelings so as not to suppress them with defense or rationalization. Intellectuals are good at concentration or focused activity. So concentration meditation (for example, repeating a mantra mentally) comes naturally to intellectuals. To be aware of their mind-brain *dosha* pattern, they must additionally practice relaxed witnessing—allow everything to come into the inner awareness without judgment, just as a juror is supposed to do with courtroom evidence.

How does one work with excessive hyperactivity? The basic objective here is to slow down. What does "slow down" accomplish?

Do an experiment. Take a coffee break right now while you are reading this book. No hurry, the book is not going to run away. Make the coffee (or tea) as a ritual, paying attention to every step.

When the coffee is prepared, sit down with a cup. Slowly lift the cup to your mouth and take a sip. Watch the response. "Ahhh . . ." You feel relaxed; you feel happy.

It is easy to rationalize away the happiness, identifying it with liking coffee. But a little experimentation will easily convince you that happiness is not inherent in the coffee but instead came from the momentary expansion of your consciousness. Slowing down, first and foremost, is a way to expand consciousness that produces happiness and bliss.

Now you can see what hyperactivity is depriving you of—bliss. The more you indulge in hyperactivity, the more it robs you of bliss. First comes sleeplessness. Sleep is bliss, being unbroken consciousness. Then come relationship problems—more separateness and less bliss. Finally, the disease—separateness has risen to a maximum. Slowing down, just by itself, makes room for the dissolution of separateness.

In 1991, I was at a yoga conference in India to give an invited talk on consciousness and quantum physics and was taking myself a bit too seriously. Then one of the teachers there asked me, "What do you do when you are by yourself?" And my psychological inflation came crashing down. I had to admit to myself that when I was by myself, I was fidgety and bored, always trying to find something to do. I realized that I needed to slow down.

How does one slow down? You can only take so many coffee breaks in a day. The primary answer here is also meditation, but the approach to meditation is different.

Hyperactivity in children is common in this country, and such children often suffer from attention deficit disorder. This is, of course, when hyperactivity is already pathological, but attention deficit is a common associate of hyperactivity even for adults. So hyperactives must learn to focus their attention, which is the objective of usual forms of meditation called concentration meditation, such as mentally repeating a mantra over and over as in Transcendental Meditation. It helps initially to learn to concentrate on other objects as well, such as on a candle flame, on the breath, and so forth.

Meditation in this concentration form has now become

famous as causing the "relaxation response," thanks to Herbert Benson's pioneering research.

After a while of practicing concentration meditation, you will realize that holding concentration for prolonged periods of time is difficult if not impossible.

Do a little practice. Sit comfortably, close your eyes, breathe evenly, and repeat a common mantra such as the Sanskrit word "Om" in your mind. Of course mind will find distraction, but as soon as you discover your distraction, bring your mind firmly back to the mantra. Do it for five minutes.

Now open your eyes. How many times did you move away from your mantra? Five times? Twenty-five times? It is hard, isn't it? It is a lot of work. And it takes a lot of practice to quiet the mind enough to hold attention for a time.

So we discover a better way to hold attention—the way of relaxation, awareness meditation. This awareness meditation is the same as I described as an antidote to excessive intellectualism.

Yoga

There are sophisticated and subtle techniques that deal with the root of the mind-brain *doshas,* the mental *gunas* themselves. Remember that the mind-brain *doshas* are produced by the unbalanced application of the mental *gunas,* the mental qualities we are born with. Intellectualism is the waste product of unbalanced use of *sattva*—fundamental creativity. Hyperactivity is the result of unbalanced use of *rajas*—situational creativity and problem-solving. If we can balance the qualities of *sattva* and *rajas* (and also *tamas*—inertia—because we do not use *tamas* enough when *sattva* and *rajas* dominate our persona), then the mind-brain *doshas* will no more haunt us.

In principle it is easy. People of *sattva* have to engage more in problems of the ordinary world and ordinary living, which require only rajasic skills. People of *rajas* have to be more interested in fundamental creativity, in the contexts of thinking itself, in the archetypal domain of love, beauty, and justice. Both kinds of people, sattvic and rajasic, must practice the thing with relaxation, balancing *tamas* in their life.

In practice, these balancing tricks constitute the heart of what is called yoga in India. The Sanskrit word *yoga* means "union" or "integration." The objective of yoga is to integrate the separate self, ego, and the universal unity called the quantum self. But why involve the quantum self in healing?

If you are not a materialist, you will be wont to use the forces of the mind (meaning consciousness) to induce mind-body healing. Consciousness has only one force, one way of manifesting its purposiveness in the world, and that one way is the freedom of choice, free will, for choosing from among the quantum possibilities the unique actuality of manifest experience. But this free choice is the domain of the quantum self. We have free will to heal ourselves only to the extent we are able to act from our quantum self-consciousness. Hence yoga is paramount.

Since *sattva* or fundamental creativity takes us beyond the mind, using the quality of *sattva* is already a yoga; it is called jnana (meaning wisdom) yoga. *Rajas,* on the other hand, is the tendency to use the discoveries of *sattva* to worldly purposes of empire-building, employing situational creativity and problem-solving skills. *Rajas* can be used for personal aggrandizement, which just serves the ego. If, however, the acts of *rajas* are done with selflessness for the good of the world, then these acts become yoga also. This is called karma (meaning action) yoga. In truth, karma yoga is better done in the service of love, and the yoga of cultivation of love is called bhakti (meaning love) yoga.

So the balancing act for the person of excessive *sattva* is to continue to practice jnana yoga with some karma yoga and bhakti yoga. And the balancing act for the person of *rajas* is to practice karma yoga in conjunction with jnana and bhakti yoga.

There is also another balancing that needs to be done—the balancing of the mind, the vital, and the physical. This is called raja yoga, codified by the great yogi Patanjali (Taimni 1961). Raja yoga incorporates hatha yoga (physical postures) and *pranayama* (breathing practices). Needless to mention, in the West the combination of hatha yoga and *pranayama* is what is called yoga. But the goal of raja yoga is to integrate the action of physical body, energy body, and the mental body so that the ego can integrate with the quantum self. So the beginning practices of hatha yoga and *pranayama* are complemented by practices of meditation as well.

If you have practiced any hatha yoga, your first impression may be that it is just stretching exercises. However, you are missing a point or two. First, hatha yoga postures are done slowly; so consciousness expands while doing the stretching. Second, in hatha yoga, one must pay attention to the internal goings-on, to the flow of vital energy. The second objective of hatha yoga is practiced more directly in *pranayama* or breathing practices. One becomes aware of the vital energy movements as one pays attention to the breath. Notice that paying attention to the breath also has the effect of slowing down the breath, and so slowing down the activity of our internal organs (Goswami 2003).

What do slowing down and expansion of consciousness accomplish? Slowing down means less quantum collapse; in between quantum collapses, there is now opportunity for unconscious processing, proliferation of quantum possibilities. This makes room for creative quantum leaps to new contexts. The expansion of consciousness helps the shift of our identity beyond the ego, making room for the quantum self to come into the picture.

Christian Science and Faith Healing

We can also recognize that mind gives meaning and faulty mind gives faulty meaning; faulty meaning is what ultimately produces disease. So why don't we correct this tendency of the mind to give faulty meaning?

Interestingly, the famous Mary Baker Eddy founded the Christian Science tradition of dealing with disease with precisely this objective in mind. So Christian Scientists are taught that disease is an illusion (which is true in the ultimate sense of Vedanta) and the mind (meaning consciousness) has the power to heal any disease by realizing that disease is an illusion. There are also charismatic Christians who believe in faith healing, following a passage in the Bible.

Unfortunately, although there is undoubtedly evidence of success in both Christian Science and faith healing, it is also undoubtedly true that these methods do not work more often than not. So what gives?

Also, the techniques I have discussed above, behavior modification, meditation, and yoga, work best as techniques of prevention of mind-body diseases and management of stress. They have quite limited success in actually healing serious diseases. Again, what gives?

So far we have left out the most spectacular class of mind-body healing—mind-body healing that is quantum healing, which requires one fundamental aspect of mind's quantum nature, discontinuous quantum leaps. This is the subject of the next chapter.

PART 4

The Healing Path to Supramental Intelligence

16

Quantum Healing

The most well-known phenomenon of mind-body healing is the placebo effect, mentioned before. Patients are given sugar pills by a physician under the pretense that it is serious medicine, and they are found to heal significantly better compared to a control group who are given the same sugar pills but with full knowledge as to their content. So the mental belief (or faith) a patient has in a pill and a doctor is very important for the physical healing (Benson 1996).

Placebo has been studied within science relatively recently, but there are anecdotes of its use that go back ages. I have heard of many sages of India who could heal. But strangely, they would give something for the patient to take orally, "Take this, and you will be all right." That something might be a piece of fruit or something equally irrelevant to the disease. But somehow it produced healing. Placebo?

Many conventional physicians think of any healing using a technique of alternative medicine as healing via the placebo effect. In biology, there are many human characteristics, such as consciousness, ethical behavior, or aesthetics, that biology has difficulty in explaining. Biologists promptly attribute such characteristics to the ubiquitous cause of "survival benefit." But they don't care to explain where the prerogative for survival, which is neither a physical nor a chemical property of matter, comes from. Similarly, allopathic doctors never ask, from whence the efficacy of placebo? Also, mysteriously, placebo cures, although real, are often only temporary. How so? Nobody asks.

Then there are all those reported cases of spontaneous healing that may be triggered by a variety of stimuli, medical procedures, and sometimes just plain intentions and faith (Schlitz and Lewis 1997). In science, unusual phenomena often give us more clues about the system we are dealing with. So what is the explanation of this particular unusual phenomenon?

Visualization has strong effects on the body (see chapter 15). Indeed, visualization has been used with some success for the treatment of cancer patients (Simonton et al. 1978). But visualization works for some people and doesn't work for many others, although they may be quite good at it. Why?

There seems to be a consensus now that includes even conventionalists that a loving environment may be conducive to healing. Similarly, the tangible effect of prayer by prayer groups for the healing of patients has been demonstrated so well that even many conventional health practitioners are persuaded as to the causal efficacy of prayer. So more and more, one finds attempts to create a loving environment and prayerful atmosphere even in hospitals based on conventional treatment. But seldom does a conventionalist bother to ask, Why does loving work for healing? Where does the causal efficacy of prayer come from?

Last, even conventional medical professionals accept the fact that a good doctor-patient relationship accelerates healing. If healing is a material phenomenon and objective, then this, too, is hard to comprehend.

We are missing an important ingredient of the healing that is taking place in these examples. We are missing the quantum aspect. There are a few explicitly quantum aspects of mind-body healing (see also chapter 6): the quantum leap, quantum nonlocality, downward causation, and tangled hierarchy. Until we include the quantum physics of mind-body healing, our understanding of some of its successes will be incomplete.

Mind-body disease consists of physical ailments in which the imposition of mental meaning sets up disharmony in our vital and physical bodies. So mind-body healing must involve changes in the meaning-context that the mind sets up for the malfunctioning of the vital and the physical bodies. Sometimes this change in the context of meaning processing by the mind comes about simply by reshuffling old contexts. This is when the continuous methods of mind-body medicine—self-hypnosis, visualization, meditation, and so forth—work. But sometimes, as in the cases mentioned—some cases of placebo, spontaneous healing, and healing through visualization—the contextual shift cannot happen at the level of the mind itself. In those cases, mind-body healing is a misnomer.

The contexts of mental thinking come from the supramental domain of consciousness; to change the context to a new one, we mental beings will have to leap to the supramental. This leap is a discontinuous quantum leap, and this is why this type of healing is quantum healing.

"Quantum healing" is a phrase that has already been creatively intuited, albeit in rudimentary forms, by at least two physicians, Larry Dossey and Deepak Chopra. Dossey (1989) emphasized the quantum nature of the healing of a patient by another (other-healing, as it is sometimes called), such as through prayer, as evidence of quantum nonlocality. Chopra (1989) correctly intuited the quantum nature of self-healing: that it consists of quantum leaps. I have already introduced their work in chapter 5. Here are a few additional details. Let's begin our discussion with Chopra's work.

Chopra's Quantum Leap

In the 1980s, the physician Deepak Chopra was looking for an explanation of self-healing. When asked if anybody can claim to know the cure of cancer, he said, "If a patient can promote the healing process from within, that would be *the* cure for cancer."

If this sounds like Mary Baker Eddy, for whom if the mind could discover that all disease is illusion then healing would follow, it is not an accident. Both Chopra and Baker Eddy are introducing the idea of healing as self-discovery. But Chopra went one important step further. He said in *Quantum Healing*, "Many cures that share mysterious origins—faith healing, spontaneous remissions, and the effective use of placebo, or 'dummy drugs'—also point toward a quantum leap. Why? Because in all of these instances, the faculty of inner awareness seems to have promoted a drastic jump—a quantum leap—in the healing mechanism."

Chopra introduced consciousness and quantum physics into mind-body healing in an attempt to initiate new scientific modeling of this self-healing phenomenon beyond classical physics, chemistry, and biology, which have no explanation for it. In this seminal book, *Quantum Healing*, Chopra suggested that mind-body interaction in self-healing occurs through a "quantum mechanical body" and is mediated by "bliss"—consciousness.

To emphasize once again, mind-body healing is not brain-body healing. Fundamental in mind-body healing is downward causation: A thought, an emotion, a belief initiates the healing process. But the brain's capacity for downward causation is dubious. So scientists who study mind-body healing implicitly or explicitly adopt a dualistic mind-body interaction model. Unfortunately, this model is also fraught with difficulties.

If mind and body are two separate substances, how can they interact without an intermediary? How would such an interaction be consistent with the law of conservation of energy in the material world? Hence Chopra's brilliant suggestion: The intermediary in mind-brain interaction is consciousness. How does consciousness mediate the interaction of mind and body? "Through the quantum mechanical body," says Chopra, a little vaguely, "mind-body healing is quantum healing."

The vagueness of Chopra's idea disappears when we realize that consciousness mediates mind-body interaction through the "quantum" nature of both mind and body. If mind and body are Newtonian objects of classical physics, there is no way to mediate their interaction without a major revision of known physics. But if both physical and mental objects are quantum possibilities within consciousness, then consciousness can simultaneously and nonlocally collapse the possibilities of a correlated body and mind to create the actual event of its experience.

The puzzle of mind-body healing is how a thought, a nonmaterial object, can cause the brain to make a material object, a neuropeptide molecule, for example, that will initiate a communication to the immune system or the endocrine system, eventually leading to healing. From the point of view of the new psychophysical parallelism, consciousness simultaneously recognizes and chooses the context-changing thought of your self-healing (from among all the quantum possibilities that your mind and supramental body offer) along with the brain-state that has the new neuropeptide molecule (see figure 17).

Of course, the quantum leap of creativity to the supramental is crucial here for healing. This is the idea that lifts quantum healing from being a plausible idea to a legitimate explanatory principle.

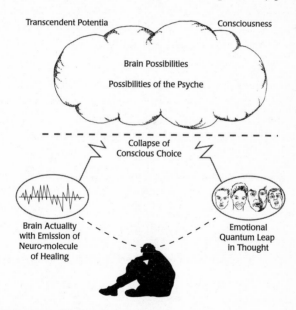

Fig. 17. How mind-body healing works.

Dossey's Nonlocality in Healing

Dossey (1989) emphasizes nonlocality as the telltale sign of the quantum and consciousness in healing. (Richard Feynman [1981] showed some time ago that a classical computer can never simulate nonlocality.) The previously discussed study by the cardiologist Randolph Byrd (1988) is one of the best to indicate the quantum nonlocality of healing. Byrd's study, involving 393 patients at San Francisco's General Hospital Cardiac Care Unit, looked at the effect of prayer carried out at a distance by several home-prayer groups.

The 393 subjects were divided into a group of 192 patients who were prayed for by four to seven different people and a control group of 201 who did not receive the benefit of any prayer at all. Neither the physician nor the patients knew who belonged to which group. Byrd found the effect of prayer, even when nonlocal, strikingly positive. For example, the prayed-for patients were five times less likely to require antibiotics and three times less prone to develop fluid in the lungs (pulmonary edema), both statistically significant results.

How does prayer work? We can easily integrate Dossey's quantum nonlocality in other-healing and Chopra's quantum leap in self-healing within one model. When somebody prays for you at a distance with purity of intention, consciousness, being nonlocal and unitive, simultaneously collapses the healing intention in your mind as in mental telepathy (although you may not ordinarily be aware of it because of noise—secondary awareness events—in your mind-brain complex). From then on, the same process, quantum healing, operates as in self-healing.

In other words, this kind of other-healing also involves self-healing. And this is true of most, but not all, cases of spiritual other-healing: The other (healer) nonlocally transfers intention at the level where choice and quantum collapse take place, the level of the quantum self or unity consciousness.

Creativity: Downward Causation in Healing

Many physicians have cited examples of spontaneous healing, some of them as dramatic as the overnight vanishing of a malignant tumor (see Chopra 1989; Weil 1995; Moss 1984). I cited two such anecdotes earlier.

As I have said before, conventionalists in the medical profession either dismiss the cases of spontaneous remission of disease under the general category of the placebo effect or, if the concept of placebo does not apply, they keep silent about such cases. Either way, they miss a grand opportunity for a new insight. Gradually, however, a new hypothesis is being considered within the medical profession—that our body already has, in many cases of illness, the requisite wisdom and mechanism for cure; we just have to discover it and manifest it (O'Regan 1987; Weil 1995). This idea, too, is limited because what is attributable only to consciousness (the power of creativity and downward causation) is attributed to the causally impotent physical body, which is mere hardware.

But suppose we boldly recognize the healing power of consciousness. Consciousness has the requisite wisdom (in its supramental compartment), the mechanism (choosing a new context for mental processing of the meaning of emotions) for cure. It also has the power to discover what is needed (the power of making the quantum leap of insight) and it has the power to manifest the insight, by unblocking vital feeling at the appropriate chakra, unblocking the vital program and thus also unblocking the correlated physical organs, which revives proper organ functions.

In truth, faith in a doctor's word as in the placebo effect gives a patient only a glimpse of his own healing capacity. To truly manifest this capacity, the entire program of creativity, going through all the stages of the creative process, which ends in no less than a change of the context of one's living, is essential.

Now the crucial question: If quantum healing involves creativity of the mind, can we develop a program of action for healing ourselves based on this idea? It is true that creativity is acausal. We used to call a creative insight God's grace. But it is also true that engaging in the creative process in its four stages (preparation, incubation, insight, and manifestation with understanding) helps creative acts. What would this entail in the case of mind-body healing?

Suppose that instead of a belief that people are getting some sort of medicine, as in the placebo effect, patients operate under the conviction (a "burning" one because of the urgency of the situation) that they already have the mechanism for healing, which they need to discover and manifest. The first step of such a creative question then is preparation.

Patients would be encouraged to research their disease (with a lot of help from their physicians, of course) and meditate on it. Such meditation will readily show the role of mind-brain *doshas* in how we deal with mental stress and how our habits of mentalization of emotions and suppression or expression of emotions, as the case may be, contribute to the disease.

One of the root causes of mental stress accumulation will also become clear: mental speed—hurrying and rushing—augmenting the pursuit of desires with accomplishments, anxieties, and daydreaming. So the purpose of the preparation stage is to slow down the mind and to create an open, receptive mind, which is an essential first step toward any creativity.

At the next stage, the patients and their doctors would try various new (to the patient) techniques of mind-body medicine. This is the stage of creativity in which we use unlearned stimuli to generate uncollapsed possibility waves of the mind and the supramental, but we don't choose among the possibilities. Since only choice can create an event of conscious awareness (Goswami 1993), what I am talking about is unconscious processing—processing without awareness.

There are well-known cases of art therapy in which people heal themselves by submersion in beautiful, spiritual art. Art therapy does not work for everybody, but how does art therapy work at all, even for some people? The mental imagination of healing inspired by the art very soon gives way to unconscious processing, opening up a new vista of possibilities. Sooner or later a seemingly inconsequential trigger precipitates the quantum leap of insight: Simultaneously, the new supramental context and the mental gestalt that represents it appear manifest in conscious awareness. The insight leads to the corrective contextual shift in how the mind handles emotions.

Manifestation of the insight begins at once: Freed from the shackles of mentalization, feelings and the vital blueprints become functional once again, leading, sometimes dramatically, to healing of the correlated organ concerned.

I have already mentioned that there are some reported successes in treating cancer patients via the use of creative visualization (Simonton et al. 1978), to which the previous scenario applies. Here is a particularly poignant description of one person's quantum healing through visualization:

> When I was in Mexico, I had started having pain in my chest. I went across the border and got an MRI scan, which showed a mass on my thymus connecting to the aorta. I decided just to wait, but a scan six months later showed it was still there.
>
> I decided to spend a week at Carl Simonton's healing center in California, and I imaged "sharks eating cancer cells" as they recommended. But toward the end of the week, I had this extremely vivid, spontaneous vision that wasn't on the program. I saw a mass on my thymus as a piece of ice that just started to melt in these big, amazing drops. I've never in my life had this kind of clear image just come up by itself. And I knew instantly the drops are just teardrops. My whole life, through all the losses, I'd never been able to cry. Now there was this melting away of the oppression I'd been feeling; the deaths and the abuse in my childhood, the unresolved relationship with my ex-husband. The emotion was suddenly available, and it felt so powerful.
>
> Four months later, I had another MRI, and the mass was gone—there was no sign of it. I had no new treatment. Whatever this mass had been, they said the only way they could tell it had ever been there was from the previous two tests (quoted in Barasch, *The Healing Path,* 1993, pp. 273–274).

Clearly, the experience released the depression of emotions accumulated through a lifetime. And there is no doubt that the experience was sudden and unexpected, a genuine quantum leap.

A spontaneous remission, in this way of looking at things, corresponds to a creative insight, when we are able to choose "the healing path" out of myriad possibilities generated by unconscious processing. This choosing is the work of unitive consciousness in its quantum self.

How does one experience this choosing of healing insight, this quantum self experience? Experiences vary. The previous example was a vision. Physician Richard Moss (1981, 1984) talks of a cancer patient who attended one of his workshops; I mentioned this one before. During the workshop, she was defiant and not responding to the various attempts of Moss to energize her. But at some point Moss broke through her shell and she responded by participating in a spontaneous dance that led to a tremendous "ah-ha" experience. The following morning the patient woke up feeling so much better that Moss felt compelled to send her for a checkup. Miracle of miracles, her cancer was gone.

The patient in Moss's anecdote experienced the "ah-ha" of creative insight. But patients also report experiencing the choice itself, when the purity of the healing intention is crystallized. As an example, here is the physician Deepak Chopra's account of the healing of a cancer patient through sudden insight:

> [A] quiet woman in her fifties came to me about ten years ago complaining of severe abdominal pains and jaundice. Believing that she was suffering from gallstones, I had her admitted for immediate surgery, but when she was opened up, it was found that she had a large malignant tumor that had spread to her liver, with scattered pockets of cancer throughout her abdominal cavity.
>
> Judging the case inoperable, her surgeons closed the incision without taking further action. Because the woman's daughter pleaded with me not to tell her mother the truth, I informed my patient that the gallstones had been successfully removed. I rationalized that her family would break the news to her in time . . .
>
> Eight months later I was astonished to see the same woman back in my office. She had returned for a routine physical exam, which revealed no jaundice, no pain, and no detectable sign of cancer. Only after another year passed did she confess anything unusual to me. She said, "Doctor, I was so sure I had cancer two years ago that when it turned out to be just gallstones, I told myself I would

never be sick another day in my life." Her cancer never returned.

This woman used no technique; she got well, it appears, through her deep seated resolve, and that was good enough. . . . I must call it a quantum event, because of the fundamental transformation that went deeper than organs, tissues, cells, or even DNA, directly to the source of body's existence in time and space (Chopra 1989, pp. 102–103).

I have cited several cases of spontaneous healing of cancer and have claimed that in each case the cause is a quantum insight. To see clearly the dynamic role that the insight plays, it may help to probe a little deeper into what has to be involved in these kinds of cases of cancer cure (Weil 1995).

There is always pressure on the cells of our body to become malignant, a condition in which they do not die at the expected time, do not stay in the same place, and in general do not conform to cellular laws of regular behavior. But malignant cells do not represent cancer, only seeds of cancer.

This is so because malignant cells distinguish themselves by displaying abnormal antigens ("not me") on their surface membranes. So the immune system, whose job is to distinguish "me and not me" can recognize them and get rid of them. In this way, cancer becomes reality only when for some reason this normal immune system function is inadequate (due to a physical or a vital defect) or suppressed (due to the mind-brain *doshas*). For example, when excessive mentalization and intellectualism lead to the suppression of feelings at the heart chakra, this also suppresses the immune system programs that are correlated with the feelings (see chapter 14).

So the spontaneous healing of cancer must be due to the sudden onset of such a dynamic surge in immune system activity that the cancerous growth is gotten rid of within days, even hours. Suppose the immune system inadequacy or suppression is due to faulty mental processing—excessive mentalization and intellectualism taking its toll. A quantum leap to the supramental is accompanied by a shift in the processing of mental meaning; this frees the blockage of feelings at the heart chakra. This then can have the desired dynamic effect on the immune system in the form of reactivating its vital program of getting rid of cancerous cells with such vigor as to effect very rapid healing.

What do the data say on spontaneous remission of cancer? The Institute of Noetic Sciences researcher Brendan O'Regan (O'Regan and Hirshberg 1993), who did perhaps the most extensive research on the subject, talked about three kinds of spontaneous remission cases: (1) pure remission—remission with no allopathic treatment after the diagnosis is made; (2) remission with some treatment after diagnosis, but the treatment is clearly unsuccessful; and (3) the most unusual kind of remission, in which the "cures are sudden, complete, and without medical treatment," associated with spiritual cures.

This third class passes as quantum healing with clear discontinuous "spiritual" experience (or insight). For the other two classes, healing may be due to situational reshuffling of previously known programs of the mind leading to adaptation of the new situation and healing. But the healing may also be due to the same kind of discontinuous quantum leap as the patients of class three, except that the participants were not observant enough to note the special-ness of the creative moment. I think that this was due to the lack of preparation; the insight was not particularly meaningful to them so they did not take notice.

The final stage of the creative process (Goswami 1996, 1999)—manifestation—is also important to discuss in this creativity model of quantum healing. Manifestation is not complete with only the reactivation of vital blueprints that are needed for the normal functioning of the organ(s) involved. After the remission has taken place, the patient has to bring to manifestation some of the lifestyle changes that are commensurate with the shift of context in the mental processing of feelings if the remission is to be stable and permanent. For example, a lifestyle that produces excessive intellectualism and defensive reactions must give way to a more balanced one.

Why do most placebo healing cases appear to be only temporary healing? I think those are not cases of genuine quantum healing by the self. Instead, the faith that "I am getting medicine from my doctor whom I trust" leads to a reshuffling of known meaning contexts of the mind that temporarily allows the mind to adapt. In other words, they are examples of healing due to situational creativity taking place spontaneously without the participation of the conscious self.

Let's again discuss the case of the former *Saturday Review* editor Norman Cousins (1989) who has written about his self-healing from a condition called ankylosing spondylitis, a degenerative disease that causes the connective tissue in the spine to wither away. Experts estimated that his chance of recovery was one in five hundred. He stopped standard medication and substituted it with megadoses of vitamin C, all this in full consultation with his physician. Rumor has it that he may also have used homeopathy. But also important, he watched funny movies (for example, old W. C. Fields flicks and the Marx brothers' escapades) and read his favorite comic books. And miraculously, Cousins completely recovered from his condition and resumed his very productive life.

I think that Cousins went from a serious disease to healing more or less following the stages of the creative process. The first stage, his hobnobbing with standard medicine and all that, was preparation. The second stage, watching movies and reading the funnies, allowed him the all-important relaxation "being" mode of creativity alternating with the "doing" mode of taking vitamin C ("do-be-do-be-do"). Eventually, he got his quantum leap, leading to recovery, and from all accounts, he did make lifestyle changes.

Very importantly, creative healing is an idea that can be medically tested. We can clinically study three groups of patients and compare their healing rates:

1. A conventional placebo group in which the patients will be given a sugar syrup or some such placebo by a doctor to stimulate a belief.

2. A creative healing group in which the patients will be aware and will be carrying out their own creative process in close cooperation with a doctor in the preparation stage (as Cousins did).

This group will also carry out the manifestation stage if a creative healing takes place.

3. A control group, which will use placebos with full knowledge but will not engage in the creative process for healing.

Tangled Hierarchy

A physician (of conventional medicine, of course) goes to Heaven and finds a big line at the pearly gates. He is not used to waiting in line, so he goes straight to Saint Peter, the officer in charge of admission. Upon hearing his complaint, Saint Peter shakes his head, "Sorry, Doc. In Heaven, even doctors have to wait to get in." But just then, one fellow in a white physician's robe goes running through the gate, paying no heed to the line.

"Ha," says our doctor. "There goes a doctor without waiting in line! How do you explain that?"

"Oh," chuckles Saint Peter. "That's God. Sometimes He thinks He is a doctor."

The point I want to make is that the role of the physician in creative healing has to change drastically. In conventional medicine the hierarchy imposed by the physician in the doctor-patient relationship is clearly a simple one: The physicians tend to think they are God, hierarchically superior to their patients who don't know anything about health and healing. At a recent conference on the philosophy of health held at the University of Oregon (Evans 2003), a family physician said, "The typical doctor-patient interaction is for the doctor to ask, 'How are you?' and the patient to respond, 'I was hoping you would tell me that.'"

But this attitude is clearly not well founded except maybe at the material level of the physical body; patients have the best knowledge of what is happening to their subtle bodies.

Actually the doctor-patient relationship is anything but a simple hierarchy; it is tangled beyond belief. I will repeat an anecdote (Locke and Colligan 1986). A doctor was treating an asthma patient who had difficulty breathing. Naturally, when the doctor heard of a new medicine, he called the company for a sample, got it, and gave it to his patient. His patient got relief in his breathing within minutes; even his bronchial tubes seemed to remain open longer.

Out of curiosity to check the efficacy of the drug, the doctor then administered a placebo to his patient. But now the patient's difficulty in breathing returned. So the doctor was convinced that the medicine worked and wrote the pharmaceutical company for more samples. Imagine his surprise when the pharmaceutical company admitted that he was sent a placebo by mistake the first time. So what explains the efficacy of the so-called medicine then? Obviously, the *doctor's* belief in the medicine.

Simple hierarchy is detrimental to creativity. If the doctors are authoritarian, their patients will not feel encouraged to think creatively about their situation. Thus in creative medicine, the now-prevalent doctor-patient simple hierarchy has to give way to a co-learning relationship—a tangled hierarchy.

Further, conventional practitioners of medicine have developed a habit of looking at healing as an objective science. However, healing is art as well as science, subjective as well as objective. You can learn all there is to learn about standing waves on the guitar string and all the rest of the physics of that musical instrument, but that knowledge will not replace the art of playing the guitar, which requires creativity of the player. Creative healing, above all, demands creativity in the physician-patient relationship, and that creativity begins with a tangled hierarchy—a hierarchical circularity of levels in which each level affects the others ad infinitum.

One of the most desirable aspects of the paradigm shift we are witnessing in medicine is that a transition from simple hierarchy to a tangled hierarchy in doctor-patient relationships is already taking place. I will illustrate this with psychologist Arnold Mindell's story of how he discovered his concept of the dream body—a person's total real personality as manifesting in different channels—while working with a terminal patient with stomach cancer.

During one of the interactive sessions, the patient had a creative insight that he wanted to "explode" in self-expression as never before. Just before the patient was to go to the hospital, he had a dream that he shared with Mindell. In his dream, he was a patient with an incurable disease that could only be treated by a medicine that acted like a bomb. Suddenly, Mindell had his own insight; he saw the underlying unity in the concept of the dream body of the patient's cancer, the bomb of his dream, and his need to explode in expression.

The creative experiences of the doctor and the patient did not end with just realizations, but they both completed the manifestation stage also. The patient left the hospital alive and stayed alive for a few years manifesting a change of lifestyle with his newly discovered expressive ability. And Mindell became famous for his successful dream-body work with patients.

Disease and Healing as Opportunities for Waking to Supramental Intelligence

Intelligence is the ability to respond appropriately to a situation. For example, the intelligence that is measured by the IQ test is our problem-solving capacity, assumed to be mental and thus algorithmic, logical, and quantifiable. Is there intelligence aside from this mental intelligence?

One other kind of intelligence touted today is emotional intelligence, thanks to a popular exposition by Daniel Goleman (1995). When we are faced with an emotional situation, mental problem-solving capacity is not of much help.

So what is emotional intelligence? The psychologist Peter Salovey (Salovey and Mayer 1990) defines it as capacities in five different domains of experience: knowing oneself (awareness of one's own emotional nature); emotion management; controlling emotions in the service of motivation toward goals; empathy (the ability to interact with other people's emotion yet retain one's objectivity); and handling emotional relationships.

The perceptive reader will already note that many of the techniques of mind-body medicine of a previous chapter (see chapter 15) are designed to help us grow emotional intelligence—for example, awareness and empathy training and chakra psychology. Thus emotional intelligence is an essential ingredient of good health maintenance and disease control.

How does emotional intelligence grow with awareness and chakra psychology? Through awareness training, we learn to feel our own emotions, discover the chakras, and develop the ability to move vital energy in and out of our chakras using imagination, vital energy massage, and so forth (see chapters 11 and 15). Through empathy training, we learn to experience the emotions that interaction with an emotionally disturbed person (nonlocally) brings us, without identifying with them. With chakra psychology, we learn to stop mentally amplifying the vital energy expressions of the vital body. Such training then enables us not only to motivate ourselves to goals and empathize with others, but also to exhibit a considerable number of relationship skills.

We can also see why emotional intelligence helps us deal better with mind-body diseases than when we employ mental intelligence alone. Whereas mental intelligence tends to lead us to a mentalization and mental creation of emotion, emotional intelligence helps us get rid of some of the evils of mentalization and mental creation of emotions.

Yet both mental intelligence and emotional intelligence are developed and applied by continuous methods. They cannot heal when the mind gets bogged down with a serious contextual crisis, and its habit of mentalizing feelings causes, first, imbalance of vital energy and, eventually, a physical disease. Let's face it. The techniques of mind-body medicine are fundamentally coping mechanisms; they help you to keep control of a bad situation. But they cannot transform the mind; they cannot change the mind's habit of mentalizing and fantasizing feelings.

Consider an example, the case of a heart disease. Due to environmental stress and your lifestyle situation, feelings arise in your navel chakra, you mentalize them, and now you have emotions of anger and irritation. As the mental habit becomes chronic, the energy flows into the navel chakra from the heart chakra that is now depleted, and you have the emotion of hostility. Chronic hostility with your intimates causes havoc with the vital energy movements associated with the physical heart. A faulty vital blueprint leads to faulty representation at the physical level, and you end up with heart disease.

Now you engage in meditation and that gives you a relaxation response. Since you are learning to be emotionally intelligent, you visualize peace to compensate for your hostility. And these things will certainly help to keep your heart problems in control. But will they cure them? No way. Your habit of mentalization will take over whenever the stimulus is strong enough, and you will end up with a heart attack.

Consider another example. You are uneasy about romantic feelings in your heart chakra, the beginning of mentalization. You don't know what to do with romantic feelings, so you begin to suppress them. The vital energies associated with your immune system are thus suppressed; eventually, this suppresses the immune system at the physical level. When the immune system fails to do its normal job of getting rid of abnormally growing cells, you get cancer.

Now you practice techniques of mind-body medicine—you begin visualizing a healthy immune system and all that. But you don't heal. What to do?

Consider still another example. Many men experience an enlargement of their prostate gland when they become older. This is a nuisance because it forces one to make several trips to the bathroom at night, interfering with sleep. If it is a mind-body disease, how does it come about? At older age, some men fantasize too much about sex, have lustful thoughts, and all that. This produces too much vital energy at the sex chakra; this gives rise to too much of the hormone testosterone, which fosters the enlargement of the prostate.

Now suppose these men try mental and emotional intelligence to curb their lustful fantasies. Can they do it? Most would, if they could. But it is not easy. Why?

The solution to a problem often lies beyond the level of the problem. The problems here—hostility, lack of love, and lust— have but one solution, love, unconditional love.

I had a friend who, when he reached the ripe age of 60, began to prominently display on his work desk pictures of *Playboy* centerfolds. When many of his visitors objected, he displayed another sign: Dirty old men need love too. He got that one right.

But unconditional love is not a mental thing; it is not even a feeling energy. Instead, love is a context, an archetype, on which many of our thoughts and feelings are based.

Where does love reside? Beyond the vital body and the mind, it is an element of the supramental body. Developing unconditional love requires a quantum leap from the vital-mental into the supramental. Love is a prime signature of supramental intelligence.

More formally, what is supramental intelligence? The supramental domain of consciousness contains the laws and archetypal contexts of physical, vital, and mental movements. When mental movement has gone unbalanced and reshuffling of old, learned contexts is unable to change a mental habit, it is time to let go of the mind and leap for the supramental. When vital energy movement is similarly unbalanced and the vital blueprint is faulty, it is time to leap into the supramental and create a new blueprint of the desired vital function. The supramental has the archetype for it. Supramental intelligence is intelligence that enables us to make these occasional forays to the supramental as needed.

In the past we have much misunderstood things. For example, when physician Walter Cannon talked about "wisdom of the body," I think he meant this supramental intelligence I am introducing. Andrew Weil, likewise, calls the body's healing system an "innate potential for maintaining health and overcoming illness." Maintaining health is a characteristic of our conditioned systems, the body-energy body-mind trio, but overcoming illness is another matter. It may require stepping out of the conditioned systems. It may require supramental intelligence.

Quantum healing, which we discussed in the previous chapter, is a doorway to supramental intelligence. The cases of spontaneous healing that have been reported are mostly examples of unexpected quantum leaps; they happened without any process behind them. This is why I call them an open doorway. When one engages in a creative process or engages in the exploration of love within a tangled hierarchical relationship to foster the quantum leaps to the supramental, then one is no longer at the doorway. Then one has entered the domain of supramental intelligence. And when these quantum leaps occur in an easy sort of way without effort, as appropriate, one is established in supramental intelligence.

Disease as an Opportunity

I previously mentioned that many mind-body healers think that disease is a creation of the patient. "What do you gain by creating your disease?" is their favorite question to their patient. This kind of question only confuses patients and makes them feel guilty.

Yet the mind-body healer is seeing an opportunity here that the patient needs to see if he is ready for it. The correct question is this: Now that you have the disease, instead of giving it a negative meaning, can you give it a positive meaning? Suppose you take responsibility for the disease and ask: Why did I create this disease for myself? What do I want to learn from it?"

The Chinese ideogram that stands for crisis means both danger and opportunity. In disease, most of us only see the danger— the danger of suffering, maybe even the danger of death. Suppose instead that you see it also as an opportunity to probe deeper into yourself, into your supramental domain of consciousness.

A disease is an expression of enormous incongruence. If the disease is at the physical level, an injury, for example, the physical representation of the injured organ is incongruent with its vital blueprint and negates the feeling of vitality at that organ. This creates an incongruence that we experience as illness. If the disease originates at the mental level because of a mentalization of feeling, the incongruence will be at all levels—mental, vital, and physical. We think something, we feel something else, and we act in still another way.

Years ago, a news reporter was doing a piece on Gandhi for which he had to attend several of his lectures. The newsperson was impressed that Gandhi did not consult any notes while giving his lectures, so he asked Mrs. Gandhi about that. Mrs. Gandhi said, "Well, us ordinary folks think one thing, say another, and do a third—but for Gandhiji they are all the same." Gandhi was congruent as to thought, speech, and action.

How do we reestablish congruence so that the mind, the vital energies, and the physical representations act in congruence? The answer is supramental intelligence.

A mind-body disease is a fantastic opportunity, a very loud wake-up call, to awaken to our supramental intelligence. It is like being hit by a two-by-four, but it is supremely effective. Yet, so far, very few people have successfully used this intelligence.

Uma Goswami sometimes works with Swami Vishnuprakasha-nanda of Rishikesh in India. Swamiji was a renunciate searching for God realization when he fell so ill in the gastrointestinal system that he could not eat anything for 29 days. An intuition told him to go and lie down at the Anant Padmanava temple in the South Indian city of Trivandram, so he did. Suddenly, he was blessed with a vision, a quantum leap to the supramental; he was healed, and his context of living changed forever. His fear of death was transcended forever.

He now lives at least part-time in what the great Indian sage Sri Aurobindo called the "intuitive mind." This is the suprarational, supramental way of living where you wait to hear your intuition before you act on anything that is nontrivial (Goswami, 2003).

When we engage in supramental intelligence in the acts of creativity, we can use a quantum leap of creative insight to the service of outer creativity, or we can explore ourselves in inner creativity (Goswami 1999). In the same way, if we are only interested in supramental intelligence to heal our disease, it is like engaging in outer creativity. That's good, but we are limiting the application.

It is entirely possible to use the search for supramental intelligence in the creativity of the vital-physical domain for the objective of spiritual growth. Then it is like inner creativity—it is great. Read Bernie Siegel's book *Peace, Love, and Healing* for many anecdotes about exceptional people who followed this path from disease to healing and then to wholeness.

There was a rabbi in a village who was a devotee of God, always talking about God's Grace. One day a flood began to rise in the local river. So one of the rabbi's neighbors came and warned the rabbi of the impending flood. "Rabbi, why don't you come with us?" he pleaded. "Don't worry. God's Grace will save me," said the rabbi. The neighbor shook his head and went away.

The flood came and the level of water reached the veranda of the rabbi's house. Another neighbor of the rabbi came in a boat and asked the rabbi to join him. The rabbi declined. "God's Grace will come and save me," he said.

The river continued to rise and now it engulfed the rabbi's entire house except the roof, so that's where the rabbi took shelter. The sheriff of the village sent a helicopter to rescue the rabbi. The rabbi was adamant. "God's Grace will find me."

So the rabbi drowned. When in Heaven, he went directly to God and asked him with not a little emotion, "God, I've loved you all my life. Where was your Grace when I needed it?"

God replied, "I sent you my Grace three times. First, in the form of a car, then in the form of a boat, and then again, in the form of a helicopter. But you didn't see it."

This rabbi had to die before he could see Grace. For some of us it is necessary to be sick in order to hear the call of Grace. If that is the case, we have to start with disease to explore the supramental. If the disease originates at the mental level, and conventional mind-body medicine techniques are unable to control it, we can die like the rabbi in the story told here, or engage in mind-body healing in search of a supramental healing insight as outlined in the previous chapter.

In the present chapter, I take up the case in which the disease is at the vital level and show how we can use disease at this level to wake up to supramental intelligence.

For many of us, it is not necessary to be sick before we heed the call of supramental intelligence. We can start with health and creatively explore the vital-physical. There is a spiritual tradition in India and Tibet that is based on this idea. I am, of course, talking about tantra. The martial arts developed in China and Japan have a similar objective (see later).

Vital Body Disease and Quantum Healing

Once the vital functions are built into the physical body hardware, the organs, we forget the supramental contexts (the contents of the vital functions) and the vital blueprints that are needed to make the programmed organs and keep them running. When we deal with the conditioned programmed movements of a living organ, we can even afford to forget consciousness, the programmer. But when something goes wrong with a program, what then? As an ongoing example, keep in mind the case of the immune cell program (killing abnormal cells that cannot stop replicating themselves) going awry, causing cancer.

We need to realize three underlying causes for organ dysfunction. The cause could be at the mental level. For example, the mental suppression of feelings at the heart chakra will cause the suppression of the immune system program and cause cancer. This we have already discussed. The cause could also be at the physical level, a defect of the representation-making genetic apparatus of the body. This we will take up later.

The third possibility is that the vital blueprints, in our example, of the immune system programs, no longer work, because the contextual environment of the physical body has changed. This we cannot fix by the techniques of vital body medicine (outlined in part 2) because of the contextual leap involved. We have to invoke new vital blueprints for the same vital functions for coping with the new context. But for this we need the guidance of the supramental.

So we need to make a quantum leap from the vital directly to the supramental, bypassing the mind. The supramental is the reservoir of the laws of vital movement and vital functions. There is a probability distribution full of vital blueprints that consciousness can use to make a representation of the same vital function. We use the quantum leap of creativity to the supramental to choose a new vital blueprint to make form that fits the new context. This new vital blueprint then enables the creation of new programs to run the physical organ level or even the rebuilding of the organ itself to carry out the required vital function.

Now the crucial question. If quantum healing involves creativity of the vital body, can we develop a program of action for healing ourselves based on this idea? What would the creative process entail in the case of the creativity of a diseased vital body that will take it from disease to healing?

One problem is that few people today have access to their vital body movements, let alone taking quantum leaps from the vital into the supramental. So preparation is needed, perhaps even more rigorous than in mind-body healing.

The purpose of the preparation stage is to develop a purity of intention of healing (a burning question at the vital feeling level), to slow down the vital body, which has to heal, and to create an openness and receptivity toward feelings. There are techniques of slowing down vital energy flow—*pranayama* exercises developed in India and tai chi movements developed in China are examples. How do we work to open up at the feeling level of our being?

Through intimate relationships. Burning questions will follow when we pursue relationships with utter honesty. This may involve allowing your partner to express feeling freely. Remember the movie *The Stepford Wives,* in which husbands made their wives into conditioned robots so that they would be compliant. The fact is, in Western culture, both men and women do this to their mates (women to a lesser extent) in the emotional arena. To do the opposite is a challenge.

At the next stage, the patients and their doctors would try various new (new to the patient) techniques of vital body medicine— acupuncture, chakra medicine, homeopathy, and so forth. This is the stage of unconscious processing in which we use unlearned stimuli to generate uncollapsed possibility waves at the vital and supramental (which guides the vital) levels; but we, in our ego, don't have the ability to choose among the possibilities.

So we wait for supramental intelligence to descend and create the same kind of revolution at the feeling level as the creative insight at the mental level does for mental thinking. The net effect of the quantum leap, the revolution, will be the coming into existence of new vital blueprints to help consciousness rebuild the diseased organ and programs for its carrying out of the vital functions. Since our feelings are related to the functioning of the programs that run the organs, as the vital blueprints begin to run smoothly, there will be an unblocking of the feeling at the appropriate chakra corresponding to what was once the diseased organ.

This unblocking of feeling at a chakra comes with such force that it is called the opening of a chakra. For example, if cancer at the vital level is healed in this way, the heart chakra will open. And, indeed, this is like the *samadhi* or "ah-ha" experience of inner or outer (mental) creativity. It is transformative. If the heart chakra opens this way, your heart is open not only for romantic love, but to universal compassion.

The final stage of the creative process is manifestation. As in mind-body healing, manifestation is not complete with only the rebuilding of the physical representation (software) needed for proper functioning of the organ(s) involved. After the remission has taken place, the patient has to try to bring to manifestation the transformative universal compassion toward all. Otherwise, the heart energy will contract once again, with disastrous consequences. In other words, when the supramental heeds your call and teaches you a new trick, you take the lesson seriously and try to live it as much as possible.

Similarly, quantum healing of the vital level of diseases at any chakra opens that chakra, and egoic expressions of feelings are transformed into universal expressions (see chapter 11). When we creatively heal a root chakra disease, our feelings of competitiveness and fear transform into confident friendliness and courage. Quantum healing of a sex chakra disease transforms the energies of sexuality and lust into respect for the self and others. In the same way, quantum healing at the navel lifts us from false pride and unworthiness to true self-worth.

At the throat chakra, quantum healing transforms the feelings of frustration and egoic freedom of speech to real freedom of self-expression. Quantum healing of the brow chakra transforms egoic confusion and ordinary clarity into intuitive supramental understanding. Finally, if a crown chakra disease is healed by a quantum leap, the leap will take us from the usual crown chakra feelings of despair and satisfaction to spiritual joy.

Creativity of the Vital-Physical Body for the Well Person

What is creativity of the vital-physical body for a well person? Recall once again that the chakras are places in the physical where vital body plans (morphogenetic fields) are represented in the physical as organs. These are the places where we feel vital energy movements associated with the programs that run the functions of the important organs of our body.

Of course, we identify with these movements as they become conditioned in our vital being, giving us a vital persona. Or rather we should say personae, plural, because at each chakra, we have a vital ego-persona, associated with our habit patterns of feeling there. Creativity of the vital-physical body for a well person is creative movement of vital energy beyond the conditioned movements of the vital physical ego/persona.

How does a well person engage with the awakening of supramental intelligence using the creativity of the vital-physical body, the creative process, and feeling as the vehicle (as opposed to thinking)?

I have elsewhere described the creative process involved in attaining the superconscious, supramental state called *samadhi* (Goswami 2000). The creative process, as elucidated before, consists of four stages: preparation, incubation, insight, and manifestation. Preparation for *samadhi* consists of many disciplines of the mind, including the important practice of internalizing our life, paying attention to what is happening inside rather than squandering away all of our effort in external activities.

The next stage of preparation is learning to concentrate on a particular thought (concentration meditation, see chapter 15). Of course, very soon we realize it is impossible to concentrate for very long as our organism is not made for that. So we learn to relax and practice concentration alternating with relaxation. I sometimes call it the "do-be-do-be-do" approach to meditation.

What happens then is that we enter the domain of the preconscious, the domain of secondary awareness experiences associated with each of our conditioned thoughts. In this domain, we have greater and greater free will, and we can choose the new thought belonging to the supramental, which is a quantum self experience, when we wish to. In this experience, there is immediacy; the subject-object split is not as prominent as in ordinary thought. There is separateness, but barely. This is *samadhi*—an experience of the universal quantum self oneness, and at the same time an experience of the supramental, unconditioned being in true consciousness.

Now suppose we do this practice, but not with thoughts, but feelings. Let's work on a heart chakra feeling—romance. I am concentrating on it, at the same time being relaxed about it with or without the object of my romance. Tantra gets its name of "left-handed path" because the practitioners often engage this practice with the romantic partner in the act of sexual embrace. But it is very difficult to transcend the need for orgasm, the habitual expression of sexuality.

If we succeed in sideswiping the "downward" movement of vital energy to the second chakra, and continue looking at the energy in the heart, a time comes when we are in the preconscious; we are dancing with the quantum self of new creative expression of romance—universal romance or unconditioned love. If we stay in this dance for a while, sooner or later, we fall into the quantum self of supramental insight, a universal feeling of unconditional love.

Vital energy, as I have mentioned before, can, with a little practice, be felt as currents or tingles within the body at the chakras. This creative feeling of unconditional love is felt as a current rising from the root chakra (or a lower chakra anyway). This rising energy is called in tantra the rising of kundalini *shakti*. *Kundalini* means "coiled up," and *shakti* means "energy, vital energy."

We imagine that the energy is coiled up at the root chakra, where it stays available (the metaphor is that of the potential energy of a coiled-up spring). Once in a while, spontaneously, the potential energy transforms into kinetic energy, moving this way and that way, but those movements just add to the confusion people have about the vital energy domain. (Indeed, many people seem to suffer when their kundalini exhibits such haphazard movement; read Kason 1994, Greenwell 1995.)

The kundalini rising experience, on the other hand, is directed movement. The process seems to create a new pathway; the energy is experienced as rising along this new pathway, in a straight channel along the spine.

Confusion is created because in the tantric literature this pathway of vital energy along the spine is called *sushumna* and is assumed to exist even before the kundalini awakening experience.

It is not new at all. But, of course, this is classical physics thinking. In quantum thinking, the *sushumna nadi* is only a possible pathway until the kundalini awakening takes place. Only then can it be said to have been actualized. In this way, it is okay to say that the channel is a new creation of the kundalini awakening experience. Energy rising through this channel gives the practitioner an intense feeling of timeless universal love that has transformative value. (That is, one has the opportunity to transform if one carries through the manifestation stage of creativity.)

The tantric tradition says that if the kundalini rises in one's experience from the root chakra following a new channel along the spine all the way to the crown chakra, then the kundalini is totally awakened. Then the control of vital energy movements becomes easy without effort. This is an awakening of supramental intelligence using the vital-physical domain of experience.

Positive Health

I would now like to introduce the concept of positive health as a parallel to the concept of positive mental health introduced by the psychologist Abraham Maslow. Maslow researched mentally healthy people and found that about five percent of all people have 16 personality characteristics that ordinary people don't enjoy. Principal among these characteristics are creativity, unconditional love, environmental independence, and humor.

These are characteristics of supramental intelligence arrived through transcending the mind to the supramental. I also think that a similar study is due for physically healthy people, especially those who work with their vital energy, who have kundalini experiences, or experiences of rising *chi* in the Chinese/Japanese system of martial arts.

The psychologist Uma Goswami makes another point. She maintains that people of positive mental health radiate positive emotions such as peace. She cites the example of the great sage of India, Ramana Maharshi, in whose proximity many people have deep experiences of peace. She calls this radiant mental health (Goswami 2003).

247

I have experienced one such person of radiant mental health, the American philosopher-mystic Franklin Merrell-Wolff (Goswami 2000). In 1984, I was still searching, still groping in the dark for a solution to the question: "Does consciousness collapse the quantum possibility wave?" I intuited that consciousness is the key also to personal salvation. But I was tired, I was unhappy, and I had doubts about my search when I met Franklin at his ranch in Lone Pine, California.

He was 97 years old and refused to talk quantum physics with me, because "that gives me headaches." So I just sat with him in his garden. In a matter of days, I was amazed to hear whispers about myself in which I was described as a "delightful" physicist. When I examined myself, I found all the unhappiness gone, replaced by spiritual joy that lasted during my stay at Franklin's ranch.

I think there are people of positive and radiant health among us in whose presence we feel vitality and unexplained tingles in the body, lightness seems to permeate the body, and joy bubbles up. These states of health are within the reach of all of us, but available only if we are ready to pursue creatively the domain of vital energy of consciousness.

Wouldn't it be nice if some of our health professionals were people awakened to supramental intelligence using the path of vital-physical creativity? Wouldn't it be nice if we went from our excessive preoccupation with disease to a preoccupation with health? If we learned to look at the glass as half-full instead of half-empty? For one thing, this would contribute to eradicating the fear of death that drives our preoccupation with illness.

A Healthy Perspective for Death and Dying

In answer to the question of what one cause drives up health-care costs in America, many people respond that it is the money we spend to keep people alive in the last months of their lives. Death is not only regarded as painful and undesirable, but also essentially as an encounter with the great void, nothingness, a finale—and there is the source of the fear of death.

But a science within the primacy of consciousness tells us otherwise very quickly. Consciousness is the ground of being; it never dies. Additionally, we have the mental and vital subtle bodies out of which the personality arises from conditioning. When we look at the mental and vital conditioning, we find that this is the result of modification of the mathematics, the algorithms, that determine the probabilities associated with quantum possibilities.

The "quantum" memory of these modifications is not written anywhere local, so it can survive the local existence from one space-time to another, giving us the phenomenon popularly called reincarnation. What survives then are not "bodies," but *propensities* of using the mind and the vital body, propensities that are popularly called karma.

But why do we reincarnate? Because it takes time to awaken to supramental intelligence. It requires many permutations and combinations of vital and mental patterns (that Easterners call karma) and many quantum leaps to learn eventually the contexts that constitute supramental intelligence.

It is this karma accrued in our vital and mental bodies that explains why we are born with certain vital and mental *gunas,* which, in our growing-up process, lead to the vital-physical and mind-brain *doshas.*

So what is death in this perspective? Death is an important part of the learning journey that we are on (Kübler-Ross 1975). Death is a prolonged period of unconscious processing, the second most important stage of creativity (Goswami 2001). The evidence of this is found in near-death experiences.

Near-death experiences (NDEs) have been known about for some time. Some people who could be regarded as clinically dead, due to a cardiac arrest, for example, after being resuscitated, report numinous experiences—being out of the body, meeting a spiritual master, going through a tunnel, and so forth. How do we explain such experiences, which require a subject-object split, when a person is clinically dead? The explanation is unconscious processing.

The NDE subjects were processing possibilities unconsciously while "dead"; only after they were revived did their possibility wave collapse, and their experience take place retroactively. This retroactive collapse of an entire pathway of events leading to the current event is called in physics "delayed choice" (see Goswami 2000; Helmuth et al. 1986; Schmidt 1993). It follows that if the patients were not revived, they would have continued unconscious processing until their next birth.

Healing as Regaining Wholeness

Interestingly, the word "healing" has the same etymological root as "wholeness." This means that healing in the ultimate sense is achieving wholeness. What does this imply?

Patanjali said that all our suffering comes from ignorance, ultimately. The ultimate disease, the root disease, is the illusory thinking that we are separate from the whole, which is what Patanjali calls ignorance. To heal the disease of separateness is to realize that we are the whole, we have never been separate, that the separateness is an illusion.

After one has healed oneself thusly, one can heal others. The philosopher Ernest Holmes (1938), who founded a healing tradition called Science of Mind, knew that the healing of another does not take willpower, but the knowing of truth: "Healing is not accomplished through willpower but by knowing the Truth. This Truth is that the Spiritual Man is already perfect, no matter what the appearance may be."

However, it would be wrong to say that the realization of truth automatically heals a pathological condition of the physical body (of the realized) for which the separation (for example, structure) has enormous inertia. What the realization does is free the realized from the illusion of identity with the physical body, from the illusion of identity with any suffering, be it disease or death.

Strategy for Positive Health

In this materialist culture, when we speak of strategies for good health, we include good hygiene, good nutrition, exercise, and a regular checkup. We are really speaking of caring for the physical body. In contrast, positive health begins when we start caring also for our vital, mental, supramental, and even bliss body.

What does good hygiene for the vital or mental body mean? Just as physical hygiene tells us to avoid harmful physical environments, similarly, vital and mental hygiene must mean avoiding vital and mental pollution.

The psychologist Uma Goswami emphasizes this when she says, "Emotions are more contagious than bacteria and viruses." So we must avoid the contamination by negative emotions and, by the same token, negative thoughts as part of good hygiene for the subtle bodies.

Nutrition also must include the vital and the mental. Since fresh food (cooked and uncooked) has more vital energy than stale food, even refrigerated food, fresh food is to be preferred. A good case can be made for vegetarianism when we consider nutrition at both the physical and the vital body levels. Especially when you consider the way we manufacture meat and poultry in this country (read Robbins 1996), you have to worry about the vital energy you get from these products. Eating the meat of a fearful and unhappy animal of negative vital energy (angry beef) can only bestow you with negative vital energy: anger, lust, fear, insecurity, and competitiveness.

Nutrition of the mental means feeding ourselves good literature, good music, poetry, art—what can be called "soul food." This is no less important than regular food. Entertainment that provokes laughter and joy is to be preferred over that which makes you feel "heavy." This is the general rule of mental nutrition.

How do we exercise the vital and mental? Here the Eastern traditions have contributed much toward the exercises of the vital body. Hatha yoga postures and breathing exercises (*pranayama*) have come from India, tai chi from China, and aikido from Japan. But as Uma Goswami emphasizes, do not engage in these exercises with hurry in your mind. Relax instead. Slowing down and paying attention to your inner space of vital energy is the objective.

For the mental body, the exercise is concentration—for example, mentally repeating a mantra such as *om*. You can practice it during work, or you can sit and do concentration meditation as in transcendental meditation practice (see also chapter 15). But concentration is work and it tires you out until you discover how to

251

alternate concentration with relaxation—do-be-do-be-do style. In this mode, prolonged concentration is possible without tiring out the nervous system.

And do-be-do-be-do occasionally will get you to the flow experience when you dance with the quantum self, when quantum leaps to the supramental are likely to happen. This is the exercise for the supramental body.

In my workshops, I often lead my participants in a flow meditation by following an idea that originally came from a Christian mystic named Brother Lawrence. Brother Lawrence, a simple-minded and good-hearted cook, used the practice he called "practicing the presence of God" to attain enlightenment. In my version, you begin by sitting comfortably. Do a quick body awareness exercise to bring the energy down to the body, then bring love energy to your heart. You can do it in a variety of ways.

Think of a loved one (your primary relationship) or of a revered one (for example, Jesus, Buddha, Mohammed, or Ramana Maharshi), or simply of God's love. Once you feel the energy in your heart, diffuse your attention (like you do from focused eyes to "soft" eyes). Let some of your attention go to peripheral activities going on around you—sounds, sights, even chores. Let it become a flow between your soft attention at the heart (being) and the stuff of doing at the periphery.

Imagine yourself taking a shower with a shower cap on. The water wets you everywhere, but not your hair. Similarly, the worldly chores grab your attention away from feelings at all the chakras, but never from your heart. Once you get the hang of it, you can do what Brother Lawrence did, live your life in flow.

Occasional creative quantum leaps are important also for the mental body, because only then does the mind get to process truly new meaning because of the new context involved. There is a story about the surrealist artist René Magritte. Magritte was walking along a street when a display cake at a confectioner's window sidetracked him. He went inside and asked for the cake. But when the shopkeeper was bringing out the cake in the display case, Magritte objected. "I want another one." When asked why, Magritte said, "I don't want the display cake because people have been looking at it." Likewise, it is healthier for your mind not always to process only those thoughts that everyone is processing. Hence the importance of creativity.

For the bliss body, the lazy person's exercise is sleep. But when we wake from sleep, although we feel happy, we remain the same even though we enjoyed being without the subject-object split. This is because only our habitual patterns of possibilities are available for us to process unconsciously during ordinary sleep. This changes when we learn to sleep with creativity in mind. Then states that are sleeplike can be reached, but when we wake up, we burst with inner creativity, we are transformed. This "creative sleep" is the best exercise for the bliss body.

If you are serious about positive health, don't forget checkups with persons of good positive health. In India, this is called *satsang*—to be in the company of people with little or no separateness from the whole. For a person interested in positive health, satsangs are more important than usual checkups, encounters with diagnostic machines in a doctor's office.

Miracle Healing: Creativity of the Physical?

Finally, a few comments about a very controversial subject: miracle healing—healing that seems to be in violation of even physical laws because it is truly instantaneous.

By miracle healing I don't mean *all* the cases of healing that happen at Lourdes and that are labeled by the Catholic Church as "miracle healing." People, usually from the Catholic tradition, go to Lourdes with incurable diseases, and there are many instances of healing among such people. The neuroscientist Brendan O'Regan (1997) studied these cases rigorously and concluded that they fall into the same category as spontaneous healing. And thus a majority of them are probably examples of quantum healing in the mind-body category; a few more are examples of quantum healing of the vital-physical body.

But a still smaller number may not fit in either of those two categories. Those are the cases I am talking about. Here also, there is some sort of quantum healing going on. But what does it involve?

In all cultures, there is much anecdotal evidence in favor of such healing. It is said that Jesus had this power. In India, there are many stories of the healing power of a nineteenth-century sage named Sirdhi Sai Baba that borders on the miraculous. More recently, Paramahansa Yogananda wrote (in *Autobiography of a Yogi*) about a sage named Babaji who restored all the broken bones of the body of a disciple who jumped from a cliff to prove his faith. In a more recent case of healing:

> There was a young boy in New York, 11 years old I believe . . . who collected salamanders when he was little. And a salamander—if you pluck off a leg or pluck off a hand—just grows another one. And no one ever told him that humans couldn't do that. They forgot!
>
> And so when he was about 11 years old he lost his leg up to his thigh or somewhere around there. . . . The doctor said, "It's all over." But [the boy's] belief systems weren't tied into that, and he just grew another one. It took him almost a year . . . grew a leg, started growing a foot. Last time I heard he was growing these little toes on it (quoted in Grossinger 2000).

I don't know if these episodes are true; I haven't verified them. But suppose they were. Is there any way to incorporate this kind of "miracle" healing in our scientific thinking about health?

We have to remember that supramental intellect provides the contexts of all movements—physical, mental, and vital. Thus once we develop a mastery over the supramental intelligence, we gain the easy-without-effort facility over the movements in all three arenas. In simple terms, this means the ability to control and manipulate all the three worlds—physical, mental, and vital.

One has to consider this carefully. Even a little change in a physical law can potentially change the workings of the universe. Obviously, we cannot contemplate universal changes in physical laws. Nobody can have that power. But a local change (manipulation) of the physical laws without harming anything else is certainly of no disastrous consequence. Creativity in the domain of the physical must be considered only within this very local context. True miracle healing, I think, falls into this category.

A Quantum Physicist's Guide
to Health and Healing

Reader: So what are you, as a quantum physicist, telling us about health and healing that we are not hearing from anybody else? What is quantum physics' unique message to medicine?

Quantum Physicist: Right now there is an established paradigm in medicine, one called conventional medicine, which has rather limited validity. There are also many traditions and techniques, some old, some new, under the generic label of alternative or complementary medicine, which try to fill the void that conventional medicine clearly has.

Conventional medicine is based on the materialist metaphysics that every disease can be reduced to some sort of physical malfunction, and therefore its cure also only involves fixing the material problem. The other paradigm, alternative or complementary medicine, consists of a disparate number of techniques, some new, some age-old. The age-old alternative medicine techniques have metaphysical underpinnings that squarely contradict the materialist faith of conventional medicine. Even the new techniques present anomalies and paradoxes to materialist thinking. But the major weakness of alternative medicine is the lack of metaphysical unity, which conventional medicine does have.

Quantum thinking within the primacy of consciousness is a way, maybe the only way, to develop an Integral Medicine that gives a metaphysical foundation for all of alternative medicine and more. It integrates alternative medicine and conventional medicine with a clear demarcation of their respective roles in the integrated paradigm.

R: But I am sure you are exaggerating. One of the early attempts at providing a metaphysical umbrella for alternative medicine is based on the holistic metaphysics that the whole cannot be reduced to the parts; in fact, it is greater than the parts. At the holistic level, there are many phenomena, subtle (vital) energies, mind, soul, and spirit, all with causal efficacy, that cannot be reduced to the atoms, molecules, cells, and the organs of the body. A disease is a malfunction of the whole organism and as such is best treated at the level of the whole organism. How is your science within consciousness superior to this holistic health paradigm?

QP: Defining the holistic metaphysics this way—"The whole is greater than its parts"—has become a problem, although originally it was thought to be the solution. Let me explain the second statement first.

With the success of molecular biology, the materialist paradigm looked so formidable that most scientists became convinced of the basic correctness of the idea that there is nothing but matter; everything, including subtle energies, mind, soul, and spirit, is matter. But quite a few scientists also remained convinced that these phenomena, subtle energies, mind, soul, and spirit, are important and have causal efficacy. Hence the idea of emergence as the basis of holism: The whole is greater than the parts because there are new emergent phenomena at every level of a whole that have causal efficacy and that cannot be reduced to the parts. Fritjof Capra is probably the most notable proponent of this model.

But you know, for the nonliving, there is no need to invoke holism. Atoms are aggregates of elementary particles. You can calculate everything that is experimentally known about atoms from the elementary particles and their interactions; nothing extra seems to emerge. The same is true for molecules that are made of atoms. You can predict everything about molecules from atoms and their interactions. It goes on like that.

Everything about solids can be calculated from the underlying atoms and molecules and their interactions, and so forth. It is only when we come to the living that something new seems to emerge at the aggregate level. If that is due to holism—the whole is greater than the parts—then you have a dichotomy of philosophy—materialism for the nonliving and holism for the living.

Now the dichotomy becomes a problem. Reductionism and holism are two irreconcilable metaphysics. Because of this irreconcilability of metaphysics, there would have to be two different paradigms of medicine: one, reductionist or conventional, the other holistic or alternative. What kind of holistic medicine is that? Common sense demands that there should be only one medicine.

R: Haven't we heard that before from Ken Pelletier, one of the pioneers of mind-body medicine? In his book The Best Alternative Medicine, *he quotes the materialist physicians Marcia Angell and Jerome Kassirer as follows:*

> *There cannot be two kinds of medicine—conventional and alternative. There is only medicine that has been adequately tested and medicine that has not, medicine that works, and medicine that may or may not work. Once a treatment has been tested rigorously, it no longer matters whether it was considered alternative at the outset. If it is found to be reasonably safe and effective, it will be accepted. But assertions, speculation, and testimonials do not substitute for evidence. Alternative treatments should be subject to scientific testing no less rigorous than that required for conventional treatments (Pelletier 2000, p. 50).*

In fact, Pelletier goes along with this demand of an "evidence based approach" to medicine and does his best to show that much of alternative and complementary medicine is passing this test. What do you think of this approach?"

QP: But this testing by clinical trials has disastrous consequences for the non-mainstream techniques of healing, maybe for all techniques of healing. First of all, what constitutes evidence? In the allopathic tradition, it is a double-blind clinical trial where randomly selected patients are given the same treatment, the others acting as the control group. Then the rate of healing is measured.

Is this procedure fair to alternative treatment techniques? There are difficulties. Many alternative therapies are individualized; each individual is to be treated differently, although all may have the same diagnosis of ailment. Well, suppose this problem is solved. Many alternative therapies are designed not for short-term benefits (which if they take place is a bonus) but for long-term benefits. To include long-term benefits in the protocol of clinical tests can be difficult.

Moreover, according to the tradition on which these alternative therapies are based, healing consists of healing the physical body, but also involves the healing of the nonphysical energy body (vital body), mental body, and so on. Obviously, the measurement criteria for these nonphysical bodies have to be different. And this comment applies even if you think, along with some holists, that entities such as the energy body and mind are emergent (but somewhat irreducible) entities of an underlying material substratum.

Another concern has surfaced from the experiments of the parapsychologist Marilyn Schlitz (1997) and her collaborator R. Wiseman. They demonstrated that there is a definite experimenter effect in situations where consciousness is involved. In other words, the outcome of the measurement depends on the intentionality of the experimenters. Now according to the alternative medicine practitioners, healing certainly qualifies as a phenomenon involving consciousness. But then, how should we evaluate the tests?

The most crucial reason that an evidence-based approach to medicine must be taken with a grain of salt is that this approach renders medicine to be entirely an empirical science. This is not compatible with the general trend of science. Ever since Galileo, the success of science has come from a two-pronged approach: Theoretical and experimental science are the two prongs. As Einstein said to Heisenberg, what we see depends on the theories we use to interpret our observations. There is no such thing as pure empiricism.

But suppose we give up the shortsighted approaches of emergent holism and evidence-based medicine. Suppose instead, we go back to the intuitions that established the traditions of the nonconventional techniques, intuitions that define the efficacy of these treatments with the efficacy of entities such as subtle energies, mind, soul, and spirit. This is what I have attempted to do in this book.

R: Quite. You are saying that a unified metaphysics for all medicine gives you a viable way of classifying all the prevalent medicine practices, even the conventional. In what way is this superior to other attempted classifications? For example, one author classifies all the different medicines according to Ken Wilber's (1993) fourfold matrix of consciousness: outer-individual, outer-collective, inner-individual, inner-collective (Astin 2002). For example, allopathy falls into the category of outer-collective, whereas Ayurveda is in the category of inner-individual. What do you think of that?

QP: I think well of such attempts at classification. The one you mentioned is complementary to what is proposed in this book.

I was once talking to the physician Elliot Dascher, who at the time was reading my book *Quantum Creativity* (Goswami 1999). There I showed that creativity fits into a fourfold classification schema: inner-outer and fundamental-situational. Elliot was very excited because while reading my book, it immediately occurred to him that one should classify various paradigms of medicine according to the creativity classes they belong to. Elliot thought that the models of medicine could also be divided into four classes: situational-outer (allopathy); situational-inner (Ayurveda, Chinese medicine); fundamental-outer (spiritual healing); fundamental-inner (meditation, yoga). This kind of exercise is always useful.

R: Can you answer the question "Who heals?"

QP: Yes, I can. The age-old traditions of alternative medicine are criticized and rejected by materialists because of their implicit dualism. But the psychophysical parallelism within a primacy of consciousness (which has freedom to choose creatively the healing alternative from among quantum possibilities) solves this problem.

The question of causal efficacy is defined clearly in my approach. The material, the vital, the mental, and the supramental present quantum possibilities for consciousness to choose from. Consciousness chooses from among these possibilities the actuality that is experienced with material, vital, mental, and supramental aspects.

R: But if consciousness chooses, why don't we always choose "health"? Why do we suffer from illness or disease at all?

QP: Ah, the quintessential subtlety of choice. In the 1970s, Fred Alan Wolf created the New Age slogan, "We create our own reality." He meant well but was much misunderstood. People first tried to manifest Cadillacs, following Wolf's dictum. They did not succeed too well, so they tried to manifest parking spaces for their cars for a while and be content with that!

But kidding aside, in our ordinary ego, we are ignorant and conditioned to suffer due to the ignorance of our healing power. Choice always happens from the possibilities that are offered. When we are in our ego, only the conditioned possibilities are offered with large probability, so the creative choice to wellness is easily missed.

R: So once we are ill, a quantum leap is necessary if we want to heal ourselves using self-healing only.

QP: This is so. And remember, not everyone is ready for quantum leaps. For them, medicinal systems, limited as they are, are better prescriptions. Also, creativity is harder if the disease is at the vital level rather than the mental, and it is a miracle if the disease is at the physical level only.

R: You have also mentioned purity of intentions. How does one best approach this?

QP: Yes, purity of intention is crucial for quantum leaps. Yet it is elusive because our intentions are so conflicted, so confused. However, the good news is that you can do a practice to develop it. It consists of four stages:

1. You start making the intention of healing from your ego; this is where you are. That is, you intend healing for yourself, healing from your particular disease.

2. At the second stage, let your egoic intention for healing yourself be generalized to the intention of healing all. After all, if everyone is healed, you are included, too.

3. At the third stage, let your intention become more like a prayer: Let the healing take place if it is in accordance with the movement of the whole, the universal quantum self.

4. At stage four, the prayer must pass into silence, become a meditation.

R: People in our Western culture are not at all familiar with the vital body. How do you propose we familiarize ourselves with our own vital body?

QP: This is a good question. In this materialist age, we don't pay much attention to feelings in the body such as tingling and shivers, but these are good examples of vital body movements, vital energy.

In chapter 11, I suggested a simple method for energizing vital energy in our palms: Simply rub the palms together and then separate them slightly while holding them as in the East Indian gesture of "namaste." The tingles you feel in this simple exercise are vital energy movements. If you pass your thus-energized palms over your face, for example, you will feel revitalized.

You may have heard of Therapeutic Touch pioneered in the United States by Dolores Krieger and Dora Kunz. When you pass your energized palms over a diseased organ, the organ gets a dosage of your vitality, which has healing power. This is how therapeutic touch works. I think learning how to use therapeutic touch is a great way to initiate oneself into the mysteries of the vital body.

R: *Really? Is it as simple as that?*

QP: Well, there is one more thing. You may have to have an open mind.

In the 1980s, when these things had just begun to percolate in the Western psyche, I remember getting a call from the University of Oregon psychology department to help them evaluate a guy who was claiming to demonstrate vital energy to them. I found a guy who looked pleasant, albeit a little frustrated. He was repeatedly rubbing his palms together and asking the people present (mostly skeptical behavioral psychologists) to put their hands in the space between his palms without touching them. Then he would ask, "Are you feeling anything?" to which, one by one, all the psychologists said no. Finally, my turn came. I put my hand in the space between his palms and immediately felt strong tingles. But when I said that to my psychologist colleagues, they still would not believe. They thought I was being naive.

R: *There is also a strong male-female difference in how we respond to feeling, isn't there?*

QP: Yes. Males are more mental, especially those we call intellectuals; for them, the vital energy tends to be active in the crown chakra, moving out mostly. So they suffer from apathy, or even depression. For them, the need is passion or, as Richard Moss puts it, "juicy physicality." They need to bring down the vital energy from the head to the lower chakras in the body, especially to the heart. Innately, men know that. This is the reason for the popularity of sex and violence on TV. Unfortunately, that does not do much to get the energy to the heart.

It's like a 1960s story I heard. A Westerner has opened up an ashram in the Himalayas in India and is becoming famous because of the wonderful responses he gives to visitors' questions, which seem to change people's lives. A New Yorker hears about him. Her friends urge her, "Why don't you go see him? You will gain much happiness." To this the woman says, "It's not time yet." Months go by. Friends who come back from visiting the guru remind the woman, "He has become very popular and very busy. Now you can see him only for 15 minutes and ask only three questions. Better go now and get the benefit of his wisdom." To this the woman says, "It's not time yet."

Months pass. The latest friend who returned after receiving benediction from the great guru says, "His health is deteriorating due to too much hard work, no doubt. Now he only allows you ten minutes and one question. Isn't it time for you to go see him?" The woman sighs, "Maybe so." So she packs her bags and sets out for India and somehow finds the ashram.

At the gate, the guard warns her of the latest rule. "Our master's health is not so good. You can only say to him three words. No more. Promise?"

"Promise. I don't need more than three words to say to him," says the woman.

Finally, she is shown to the master, who is sitting on a pillow. Surprising her escort, the woman does not bow. But she keeps her promise of saying only three words to the master. In a perfect New York accent, she says to the master (who is also her husband), "Irving, come home."

As this story points out, all men need to hear (and who better than a woman to tell him this?) that all their manly shenanigans in the outer world are for naught if they are not grounded in their "homes" and physical bodies.

Remember the character of the Tin Woodsman in *The Wizard of Oz*? Of all things, the Tin Woodsman desires a heart, and who is to give it to him? Ultimately, Dorothy.

R: Excellent story—funny, too. So women must be grounded in their bodies!

QP: They are. But for women also, there is a problem. Women are vital beings, mostly. The vital energy movements are generally concentrated in the lower chakras. Their energy keeps going out of the heart chakra, where the depletion is experienced as jealousy or envy.

Sri Aurobindo jokingly used to call this tendency the vampire tendency, because if you have it, you are always trying to suck out the vital energy from others to cure the negativity of the heart. You become needy. So for women, the challenge is to transform this negative energy into positive (Goswami 2003).

R: So you don't suggest that intellectuals suffering from depression should take Prozac?

QP: Hell, no. Prozac just depresses the depression. It also dulls the way we experience everything! Learning more about vital energy and learning to move it from one chakra to another is much more rewarding in the long run.

R: You mentioned The Wizard of Oz *and made a parallel of the character of the Tin Woodsman and the mind-brain* dosha *of intellectualism. There are two other characters there, the Scarecrow and the Cowardly Lion. What do they represent?*

QP: The Scarecrow looks for his "brain"; he represents the mind-brain *dosha* of mental slowness. Then of course, by default, the Cowardly Lion has to represent the *dosha* of hyperactivity. I suppose that should be clear enough. The Lion is a type A hyperactive personality, no doubt. But there is an interesting subtlety here. What is the Lion searching for?

Courage. Courage for what? The courage to create. Hyperactives need to balance their tendency for engaging too much *rajas,* for too much empire-building situational creativity, with the *guna* of *sattva*—fundamental creativity—in their lives. This will also balance their *tamas,* slowness, because without slowness one cannot be creative (see also Barasch 1993).

R: On the same subject, how do we create the sense of abundance of vitality?

QP: For that we must feel vital energy in the heart, and in general in all the chakras, lower and upper.

So first we de-mentalize our emotions. Worry is one way that the mind keeps control of vital energy. So we practice replacing worry with peace and love. For example, when the worry of money (insecurity) comes, we think of our primary love relationship (security).

Second, we engage in tangled hierarchical relationships, with circular causality instead of one-way causality. This is a surefire way of detaching from habitual patterns of both mental and vital movements.

Third, vital energy is abundant in nature. There is a tai chi practice you can do while standing in the midst of natural beauty, under open sky. Spread your arms with palms upward. Now say aloud: "Kindness and benevolence at base, frankness and friendliness at heart." Very soon you will find your palms are tingling with abundant vital energy.

R: Tell us more about de-mentalization.

QP: The mind naturally turns to dominate the vital domain of our experience by giving meaning to meaning-neutral feelings, a tendency I call mentalization. The trick is to change this habit pattern, so that the mind can turn toward the supramental instead, where it is the servant.

So we observe the ways we mentalize our feelings, our specific pattern for giving meaning to our feelings. Once we know our pattern, with utter honesty, we become open to change. And remember, changing a pattern is always done best by taking a creative quantum leap.

R: A creative quantum leap of the mind always means a shift in the context in which we process meaning, right?

QP: Right. Our contexts become so fixated, because reason works within a fixed set of beliefs—a belief system. If one of these beliefs has to change, the whole system may have to be questioned. The conditioned ego-mind hates and fears that.

The mistake we make is to think that we can change our perception of meaning just by reading something, or following a teacher, or even engaging in practice, but that is just preparation.

Have you ever been to a Zen master's teaching? He may pick up a hand fan and ask you, What is it? If you say it is a fan, he will say, I will hit you (suggesting, if you are subtle enough to comprehend it, that a fan can also be used as something to hit with). But here is the thing. If you wise up and say, it is something to hit with, the Zen master won't be satisfied. He may say something like, "Thirty percent," at best.

So what's going on? In a famous Zen book, the author describes the state to which he leapt after five days in the *zendo.* He ran to his teacher, took the fan from him, and hit him with it. He then scratched his body with it and used the fan as a scale. All these shenanigans of joyful spontaneity left no doubt in his teacher that the student was acting from his supramental quantum self, a state in which our action comes from certainty, not cleverness.

You cannot choose health just by wishing it, which is cleverness. When you choose health from certainty, after a quantum leap, only then may you be able to master the energy to make lifestyle changes. And even so, you may not; it is that complicated.

R: What is vital body karma?

QP: We bring certain propensities from our past-life experiences to this life, those we think will suit us best to fulfill our learning agenda. This is the vital karma that gives us our *doshas.* If we bring the propensity of using too much creative energy *(tejas)* in making the physical representations of the vital, we develop the *dosha* of *pitta,* and so forth.

R: But we don't try to "burn" vital karma, inherited vital propensities.

QP: Why not? After its job is done.

R: Then why not also perfectly balance the doshas *as well?*

QP: There's not much point in it. The physical body's job is to make representations, for which all we need is a body in homeostasis. Thus all we need is to keep the *doshas* near to their natural homeostasis or *prakriti* for each individual.

R: So how do we do that?

QP: Here is a simple prescription: Eat a vegetarian diet with plenty of fruits and vegetables, drink plenty of water, do everything slowly, and have a lot of relaxation time including sleep (Goswami 2003). There is a lot of wisdom in that. I would add one thing: Have creative enterprises in life (outer and inner).

R: How about those exercises you suggested earlier—physical exercise for the physical body, hatha yoga postures, breathing practices, tai chi for the vital body, concentration and awareness meditation for the mental body, flow experiences for the supramental body, and creative sleep for the bliss body?

QP: Those are exercises to keep all five of our bodies in optimal and dynamic health. Ultimately, our best strategy for living healthfully is to maintain our body in optimal health.

This is why I always say, revamping education is our most important task right now. Without our so much as knowing it, education has bought into materialism and scientism with religious fervor, and this in the name of secularism. What irony! So even six-year-olds are taught that everything is made of atoms and develop a prejudice that is so hard to overcome later on. If we are only the dance of atoms, then what choice do we have but to become confused and cynical (like those existentialists that Woody Allen portrays so beautifully)?

My hope is that the current paradigm shift will be completed soon so that we can allow our educational system to be liberal enough to teach alternatives. Give kids the choice of yoga, meditation, creativity, and emotional and supramental intelligence.

R: You have pointed out that current research has helped decipher the role of the disparate bodies of consciousness. The supramental defines the laws and archetypes of the movement of all the bodies; mind gives meaning to the vital and the physical. The vital contains the morphogenetic fields that shape living forms, and the physical, through its micro-macro division, acts as the hardware that makes software representations of the vital and the mental. What is the response of the scientific world to this?

QP: Good, as far as I can tell. But one difficulty we have for this particular paradigm shift is that it requires a multidisciplinary approach. Most scientists today are very specialized. Medicine is so specialized that it has become a joke that somebody can specialize to treat only the big toe of the right foot.

R: You say that disease can happen due to malfunction and disharmony at each of the five levels of our existence, at each of the five "bodies" of consciousness. You also say the disharmony and disease can spread from one level to another. Similarly, healing at one level can spread to other levels.

Therefore, at each level, at each of our bodies, there is the possibility of developing systems of treatments variously called material body medicine, vital body medicine, and mental body medicine. But then the vital body medicine and mental body medicine are neither alternative nor complementary; instead they define different territories of applicability depending on the origin and the amount of spreading of the disease. Isn't that so?

QP: You are very observant. For example, if the origin of the disease is a cholera bacillus and we detect it immediately, only treatment at the physical level will be necessary. Similarly, if the disease is due to excess vital energy utilization at an organ site as in the case of a stomach ulcer, then only vital body medicine is called for.

If, however, the ulcer is severe and needs immediate release, some physical body medicine as a complementary aid is welcome to give immediate relief. If the disease is due to the mind (mind-body disease), then the primary treatment must be at the mental level. Vital medicine would be a complementary secondary treatment, and physical intervention will be necessary only to deal with any urgency of the situation.

R: Can you give us some examples of when and how to apply Integral Medicine?

QP: Often intervention at the vital and mental levels has a tactical advantage. Suppose an organ is not functioning properly (irrespective of the source of the malfunctioning), causing the feeling of severe pain. We can treat the pain through narcotics. But a much safer treatment is through acupuncture. (Note that the physical effect is the same—emission of endorphins.)

Another example is mental stress, which has a potential to create disease at the physical level. We can treat stress at the physical level by taking drugs that would calm us down, but often with disastrous consequences. The mental alternative, in this case the superior, is to deal with stress through meditation and the like, interventions that have no side effects and that also deal with the problem at the source.

Admittedly, there are tricky situations such as cancer. Here we do not yet know how to determine the source of the disease. Is the disease due to a genetic defect? We can answer this question to some extent through consideration of the family history of the patient. Is the disease due to vital energy imbalances? We can use intuitive diagnosis for addressing this question to some extent. Or is the disease a product of mental stress and suppression of emotion? Which we can also address to a limited extent by looking at the lifestyle. But common sense can already dictate an integral treatment.

There is urgency with cancer: It spreads throughout the body. To prevent spreading, we must at once intervene physically through surgery and/or radiation. Then instead of chemotherapy (which has the most adverse side effects), we can administer the techniques of vital body medicine and mind-body medicine (both within a program of creativity, see chapters 16 and 17).

Similarly, with heart disease, Integral Medicine dictates that short-term physical intervention may be necessary, but long-term treatment must involve the vital and the mental levels.

Generally, we will be on sure ground if we remember the correspondence limit—emergency—for which physical level intervention is a must. For long-term treatments, the vital and the mental levels are the essential ingredients of a successful treatment; physical level intervention is optional, depending on compatibility.

R: The most important accomplishment of your approach in this book is to provide theoretical, scientific explanations of why and how the successful techniques of alternative medicine work. Ayurveda, Chinese medicine, acupuncture, chakras, mind-body healing, spiritual healing, even homeopathy, all are given satisfying explanations. But you haven't mentioned naturopathy. How come?

QP: Naturopathy, if I understand it correctly, uses all the systems that you mentioned, all healing systems available in nature. So I think naturopaths would be very happy with Integral Medicine as developed here, since we integrate all the various systems they use under one paradigmatic umbrella.

R: Do you think your Integral Medicine is the final word?

QP: Not at all, but it is a good beginning. We are starting with correct metaphysics. To be sure, the theory offered is just a skeleton and much flesh will have to come to it from further research.

R: I also like the other important accomplishment of the integrative paradigm: It enables us to consider positive health, for which the ingredient is the supramental level of healing. This is also the entry point to supramental intelligence, is it not?

QP: Yes. Even disease can be used as an entry point to the supramental, those quantum leaps of quantum healing! But it is best to approach the supramental from the side of health; it is also easier.

R: How would you evaluate health and healing in earlier days compared to now?

QP: Earlier, disease was due to a faulty representation-making capacity of the physical (genetic), environmental change of seasons, and attacks of viruses and bacteria. We had no approaches then to correct for the first, even no understanding. But vital body medicine, as in Ayurveda, Chinese medicine, and naturopathy, gave us a good way to treat seasonal effects of disharmony and most viral infections, but not such a good way to deal with lethal bacterial infections that require fast-acting interventions.

The success of modern allopathic medicine has come from the public health aspects of good hygiene, the treatment of bacterial infection (prevention through vaccination and antibiotics), and the marvels of surgery, including organ transplants. The failure of allopathic medicine as our savior came when our lifestyle became too separate from nature, so separate that proper nutrition of the vital body began to suffer. Then the stress of modern life grew to such an extent that many or most physical diseases began to have their cause elsewhere, in the vital and in the mental body.

R: Then Integral Medicine is making its appearance at the right time, would you say?

QP: Indeed so. Now with the timely development of Integral Medicine, we can remove the inadequacy of modern medicine by including vital and mental medicines with the physical as and when appropriate. We can even prescribe supramental medicine, quantum healing, for those who are ready for it.

The same thing is true about the preventive aspect of medicine. Until recently, the only preventive aspect of allopathic medicine was public health hygiene and vaccinations. But the lifestyle changes in the late twentieth century demanded a social theory of medicine, with the result of increased awareness of the harmful effects of smoking and alcoholism, and the beneficial effects of good nutrition, exercise, and so forth. Integral Medicine suggests additional possibilities of prevention by including prevention not only at the physical, but also at the vital, mental, even supramental and bliss levels of our consciousness.

R: What is the most influential finding of Integral Medicine?

QP: To bring the thrust of supramental intelligence to health and healing is of supreme importance and even of some urgency. Remember, even now, we do not have any surefire material level invention that will correct genetic defects; the inadequacy of the representation-making apparatus itself is causing disease. Materialist paradigm research has completely deciphered and codified our genetic information, but the application of that information to correcting genetic defects remains only promissory. Gene therapy is not panning out. The approach of applying supramental intelligence to this problem may be a more viable approach.

The greatest application of Integral Medicine is beyond the healing of disease. There is now the definite possibility of understanding human spirituality as a healing and to apply what we can learn from quantum healing to the healing of our spiritual selves. The social consequence of this would be far-reaching.

Ageless Body—Myth or Science?

How far can we go with positive health? Can we completely get rid of disease, even chronic little complaints? If we can totally bypass the wear and tear of the physical body, doesn't it make sense to expect that aging itself will stop?

One of the lasting myths of all cultures is the myth of the ageless body—immortality. Even materialists talk about immortality, using the promissory power of nanotechnology (Tipler 1994). Physician Deepak Chopra (1993) wrote a more reasonable book suggesting that as we grow self-knowledge, maybe we can grow long life also. Are there other ways that we can fantasize about immortality?

The connoisseurs of how we age raise the specter of the so-called Hayflick effect at this point. By doing experiments with human cell cultures in a test tube, the physician Leonard Hayflick (1965) discovered that human cells could divide only about 50 times, no more. Since, during the course of our lives, the physical body renews itself in an ongoing way via cell division, this puts a limit of roughly one hundred years to human longevity.

What then about all the stories of people of great longevity (perhaps going up to 150 years of age) that psychologist Ken Pelletier wrote about? In India, it is a common myth that there are many yogis who routinely beat the limits of the Hayflick effect. How do they do it?

Uma Goswami (2003) actually met such a yogi when she was a college student. She was visiting the southern tip of India, a beautiful place called Kanyakumary, with her parents when she heard of a 260-year-old yogi known as Mother Mayi, who, it was said, lived under water most of the time. Uma was told to wait by the seashore, which she did. Time passed and her companions all left. After about four hours of waiting, a bunch of dogs appeared from nowhere, and soon, Mother Mayi appeared from under the water. She didn't look that old, and although she didn't talk, Uma had a wonderful, joyful time with her. (Mother Mayi has since left her body, but there is a temple in Kanyakumary in her name, and the people there still remember her.)

I think people can beat the Hayflick effect to some extent because, with a slowed-down lifestyle, one can certainly slow down the need for cell division and thus stretch the Hayflick lifetime we are allowed.

But from this to immortality is a quantum leap in conceptualization. Is immortality possible?

If you have read one of my earlier books, *Physics of the Soul,* you already know that my answer to the question of immortality is a cautiously optimistic yes. My reasons are partly evidential, partly theoretical.

The evidence is iffy, no doubt, anecdotal at best. The most persistent anecdote is Jesus' resurrection. Did Jesus really revive in an immortal physical body, which a few (more than one) of his disciples saw? This seeing by more than one person is an important part of the anecdote, because only then can we say with some assurance that there was consensus viewing, and therefore the body that was seen could be physical (gross) in nature.

India, of course, abounds in such anecdotes. I will mention one, that of a sage named Babaji, who is mentioned in Paramahansa Yogananda's famous book *Autobiography of a Yogi.* Apparently, Babaji also passes the test of consensus reality. Many people (including Westerners) claim to have seen him, albeit in different times and places.

Can we make a theory to make sense of this anecdotal data suggesting immortal physical bodies? To answer this, let's delve into the mystic philosopher Sri Aurobindo's ideas about the next stage of our evolution (Aurobindo 1955, 1970, 1996).

Where Is Human Evolution Going?

I have discussed the meaning of biological evolution from a science-within-consciousness point of view before. Biological evolution is evolution of representation-making for the vital body (see chapter 6). With the development of the brain, mind can be represented, and the thrust of evolution shifts to the evolution of representation-making of the mental. This is where we are today. We live in a mental age; we thrive making more and more sophisticated representations of the mind. Witness how we use the mind even for shopping via computers, giving up the physical pleasures of touch, taste, smell, and all that.

Is there life after the mental life that we enjoy today? To see very clearly what is in store for human evolution, let's look dia-grammatically at Aurobindo's philosophy of involution and evolution (see figure 18). What is involution?

The involution of consciousness is the creation of limits so that manifestation is possible. Initially we have consciousness and all its possibilities, which include those of the past, present, and future. So no time direction can exist and no manifestation. At the first stage of involution is the first limitation: to play the game within a set of rules and archetypes. This is the supramental.

Then we have the mental limitation of playing only with those possibilities that have meaning. Then let's further limit the possi-bilities; let's play only with specific vital blueprints for making bio-logical form. This gives us the vital world. And finally, the physical arrives at the scene, the hardware with which to represent the sub-tler possibilities of consciousness. Now evolution begins, of which we have completed the first two stages already. As Pierre Teilhard de Chardin observed, the first two stages have given us the bios-phere and the noosphere. What's next?

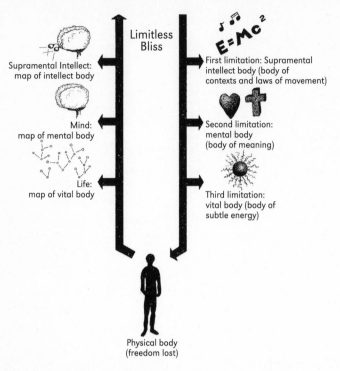

Fig. 18. Our evolutionary future according to Sri Aurobindo.

We have the supramental, of course, to represent next in form. Here is Aurobindo's great observation and prediction: The next stage of human evolution is the evolution of the capacity for making physical representations of the supramental. Once this capacity is gained, evolution will continue with the development of more and more sophisticated representations of the supramental.

But What Does It Mean to Make Maps of the Supramental?

To understand what making a physical representation or map of the supramental means, let's first analyze how we make a representation of the supramental in this mental stage of our evolution. Suppose you have a creative "aha" insight about love; that is, you take a quantum leap to the supramental and directly encounter the archetype of love. But there is no physical apparatus in your body to make a direct memory of the supramental component of that collapse experience.

So what happens is that the mind makes a temporary memory of the supramental, a mental representation. There is certainty in this experience, so your mental perspective of looking at love changes forever. However, the mind, because of quantum uncertainty, is not capable of making a permanent memory or representation. So what you do is make a brain memory of the mental representation, the meaning that your mind puts on your particular experience. But such a representation, being of a secondary nature, will always be imperfect.

So mystics tell us to live our brain representations of the mental transcript of the supramental experience with the hope that if we live our experience, with practice (practice makes perfect!), we learn the archetype.

But your mental meaning will rarely be the same as somebody else's mental meaning of an archetypal experience. So your teaching based on what you learned will be different from that of other such teachers, creating much confusion. This is what creates the different religions.

Obviously, this kind of learning is never perfect. This is the reason that most of the spiritual teachers we know today, sooner or later, get entangled with scandals. When push comes to shove, they are unable to live what they discovered.

How do people like Jesus and Buddha learn to love in its perfection? They never "learn" it perfectly; nobody can, without the power of making direct physical representation. But a Buddha or a Christ has permanent access to the supramental (a state of consciousness called *turiya* in Sanskrit), so they can always love moment to moment by directly invoking the supramental.

The ability of making a direct physical representation of supramental archetypes will change all that difficulty of learning the archetypes: love, beauty, justice, truth, good, and all that. Then we should be able to live these archetypes as easily as we learn to compute two plus two. This will also drastically change the way we teach these archetypes.

So Aurobindo called this stage of evolution the bringing down of the gods (the archetypes), down to the physical plane. Perhaps it will require a transformation of matter itself.

You may have read Jean Shinoda Bolen's two beautiful books, *Goddesses in Everywoman* and *Gods in Everyman*. Aren't gods and goddesses already represented in us? Bolen is talking about the potential of representing the archetypes, gods and goddesses in ourselves, through mental effort. But unfortunately, this way also creates shadows; and, in truth, we can never clean up the shadows, not completely.

Carl Jung's thoughts may also resonate with Aurobindo's. Like Aurobindo, Jung also saw that humanity's goal of evolution is to make the "unconscious conscious." But what Jung (1971) called the collective unconscious is what Aurobindo (and I also) calls the supramental. As to how to make it conscious, how to manifest the archetypes, Jung's preferred method was expressed by the metaphor of "alchemy." Alchemy is the idea of changing base metal into gold, in other words, the transformation of matter, the same as Aurobindo's idea.

How Does This Idea of Evolution Connect to Immortality?

The next question is: What would this representation-making apparatus be like? A superbrain, the neo-neocortex that some neurophysiologists talk about?

Think about it. We are talking about representing archetypes in our body and living them. One of the archetypes is Truth itself. What good is Truth if it is not eternal, or at least lasting for a very long time? So making representations of archetypes in this very entropy-dominated physical body is out of the question.

We have to think of physical matter that is much more subtle than ordinary gross matter, yet less subtle than the subtle bodies (vital, mental, supramental). This "supramental" matter must have micro-macro division, so tangled hierarchical quantum measurement can take place in the world of such matter. The entropy law—disorder replaces order—sits on a back burner, so the entropy arrow of time barely operates. As a result, everything seems forever new, and aging is very slow.

Another way of seeing this is that since the supramental is now mapped into the physical, its guidance of the mental, the vital, and the physical is readily available. And thus any problems with these lower bodies can be readily corrected. There is not much point of disease in such a world.

What is the relationship of this supramental matter to our ordinary matter? To the ordinary material world, the supramental matter world will be invisible, of course; that is, no direct interaction between them is allowed. The worlds will be mutually experienceable only through the intermediary of consciousness, not dissimilar to how we experience our subtle bodies.

Of course, the supramentalists will have a built-in advantage over the mentalists (us) in that they have mastery over all three levels—mental, vital, and physical, including our gross physical world. So they can go in and out of our world at will, but we will only be able to see them if they grant us temporary power to do so.

Is This Evolution Going to Happen Anytime Soon?

This is the million-dollar question. I think there are indications that we are ready for this big change.

First, the mental age has been showing signs of decay for some time. Certainly, it has gone through its prime. Today, poetry is dead. The Nobel Prize winner in literature in 2002, V. S. Naipaul, said something to the effect that novels as literature are dead, giving way to best-selling trash that looks like novels, but these books are as unreal as science fiction. Similarly, Western classical music is making more and more room for mindless light stuff.

This is only part of the story. You may have noticed that today's major problems are unsolvable through mental gymnastics. These problems include:

• problems of environmental pollution and global warming

• problems of energy shortage

• problems of nuclear warfare

• problems of maintaining democracy in the face of media power

• problems with the economics of progressivity, and last but not least,

• problems with the economics of healthcare.

One is not entirely joking if one surmises that we have to become immortal because it will otherwise be economically impossible to care for our health.

The third type of evidence is more controversial. The idea is that recently there has been an increase of visitors to us from "outer space." I am, of course, talking about the UFO phenomenon. Some of these visitors are low-level beings, suffering from negativity and ill health like us, and not interesting. But some of the visitors are radiant beings with supernatural powers from a supramental civilization. Now why would such people come to visit us unless we are ready to make the evolutionary leap to their level?

In the late 1980s, the biologist John Cairns and his collaborators (1988) made the important discovery of directed mutation: Bacteria, when starved, direct their own mutation rate to increase so that they can transform to a species that has plenty of food in the environment (see also Goswami and Todd 1997). If bacteria can do it, if bacteria can evolve to survive, so can we. And survival in this case means only one thing: to wake up in the supramental world.

Bibliography

Achterberg, J. (1985). *Imagery in Healing*. Boston: Shambhala.

Ader, R. (1981). *Psychoneuroimmunology*. New York: Academic Press.

Aspect, A., Dalibard, J., and Roger, G. (1982). "Experimental test of Bell inequalities using time-varying analyzers." *Physical Review Letters,* vol. 49, pp. 1804–06.

Astin, J. A. (2002). "An integral approach to medicine." *Alternative Therapies,* vol. 8, pp. 70–75.

Aurobindo, Sri. (1955). *The Synthesis of Yoga*. Pondicherry, India: Sri Aurobindo Ashram.

———. (1970). *Savitri*. Pondicherry, India: Sri Aurobindo Ashram.

———. (1996). *The Life Divine*. Pondicherry, India: Sri Aurobindo Ashram.

Ballentine, R. (1999). *Radical Healing*. New York: Harmony Books.

Banerji, R. B. (1994). "Beyond words." Preprint. Philadelphia, PA: Saint Joseph's University.

Barasch, M. I. (1993). *The Healing Path.* New York: Tarcher/ Putnam.

Bass, L. (1971). "The Mind of Wigner's Friend." Harmathena, no. cxii. Dublin, Ireland: Dublin University Press.

Benson, H. (1996). *Timeless Healing.* New York: Scribner.

Blood, C. (1993). "On the relation of the mathematics of quantum mechanics to the perceived physical universe and free will." Preprint. Camden, NJ: Rutgers University.

———. (2001). *Science, Sense, and Soul.* Los Angeles, CA: Renaissance Books.

Bly, R. (1977). *The Kabir Book.* Boston: Beacon.

Bohm, D. (1951). *Quantum Theory.* New York: Prentice-Hall.

Brown, D. (1977). In *International Journal of Clinical and Experimental Hypnosis,* vol. 25, pp. 236–73.

Byrd, R. C. (1988). "Positive therapeutic effects of intercessor prayer in a coronary care unit population." *Southern Medical Journal,* vol. 81, pp. 826–29.

Cairns, J., Overbaugh, J., and Miller, J. H. (1988). "The Origin of Mutants." *Nature,* vol. 335, pp. 141–45.

Chopra, D. (1989). *Quantum Healing.* New York: Bantam.

———. (1993). *Ageless Body, Timeless Mind.* New York: Harmony Books.

———. (2000). *Perfect Health.* New York: Three Rivers Press.

Cohen, S., Tyrrel, D. A. J., and Smith, A. P. (1991). "Psychological stress and susceptibility to common cold." *New England Journal of Medicine,* vol. 325, pp. 606–12.

Coulter, H. (1973). *Divided Legacy.* Washington, DC: Wehawken.

Cousins, N. (1989). *Head First: The Biology of Hope.* New York: Dutton.

Csikszentmihalyi, M. (1990). *Flow: the Psychology of Optimal Experience.* New York: Harper & Row.

Dantes, L. (1995). *Your Fantasies May Be Hazardous to Your Health.* Rockport, MA: Element.

Dossey, L. (1982). *Space, Time, and Medicine.* Boulder, CO: Shambhala.

———. (1989). *Recovering the Soul.* New York: Bantam.

———. (1991). *Meaning and Medicine.* New York: Bantam.

———. (2001). *Healing beyond the Body.* Boston: Shambhala.

Eddy, M. B. (1906). *Science and Health with Key to the Scriptures.* Boston: First Church of Christ, Scientist.

Einstein, A., Podolsky, B., and Rosen, N. (1935). "Can quantum mechanical description of physical reality be considered complete?" *Physical Review Letters,* vol. 47, pp. 777–80.

Eldredge, N., and Gould, S. J. (1972). "Punctuated equilibria: An alternative to phyletic gradualism." In *Models in Paleontology,* ed. T. J. M. Schopf. San Francisco, CA: Freeman.

Elsasser, W. M. (1981). "Principles of a new biological theory: A summary." *Journal of Theoretical Biology,* vol. 89, pp. 131–50.

———. (1982). "The other side of molecular biology." *Journal of Theoretical Biology,* vol. 96, pp. 67–76.

Evans, A. (2003). "The art and science of health." *Oregon Quarterly,* Autumn, pp. 28–32.

Feynman, R. P. (1981). "Simulating physics with computers." *International Journal of Theoretical Physics,* vol. 21, pp. 467–88.

Frawley, D. (1989). *Ayurvedic Healing.* Salt Lake City, UT: Passage Press.

———. (1996). *Ayurveda and the Mind.* Twin Lakes, WI: Lotus Press.

———. (1999). *Yoga and Ayurveda.* Twin Lakes, WI: Lotus Press.

Freedman, H. S., and Booth-Kewley, S. (1987). "Disease-prone personality." *American Psychologist,* vol. 42, pp. 539–55.

Goleman, D. (1995). *Emotional Intelligence.* New York: Bantam.

Goleman, D., and Gurin, J. (ed.) (1993). *Mind-Body Medicine.* New York: Consumer Reports Books.

Goswami, A. (1989). "The idealistic interpretation of quantum mechanics." *Physics Essays,* vol. 2, pp. 385–400.

———. (1990). "Consciousness in quantum mechanics and the mind-body problem." *Journal of Mind and Behavior,* vol. 11, pp. 75–92.

———. (1993). *The Self-Aware Universe: How Consciousness Creates the Material World.* New York: Tarcher/Putnam.

———. (1994). *Science within Consciousness.* Research Report. Sausalito, CA: Institute of Noetic Sciences.

———. (1996). "Creativity and the quantum: A unified theory of creativity." *Creativity Research Journal,* vol. 9, pp. 47–61.

———. (1997). "Consciousness and biological order: Toward a quantum theory of life and its evolution." *Integrative Physiological and Behavioral Science,* vol. 32, pp. 86–100.

———. (1999). *Quantum Creativity.* Cresskill, NJ: Hampton Press.

———. (2000). *The Visionary Window: A Quantum Physicist's Guide to Enlightenment.* Wheaton, IL: Quest Books.

———. (2001). *Physics of the Soul.* Charlottesville, VA: Hampton Roads.

———. (in press). "Quantum physics, consciousness, and a new science of healing." *Savijnanam: Journal of the Bhaktivedanta Institute.*

Goswami, A., and Todd, D. (1997). "Is there conscious choice in directed mutation, phenocopies and related phenomena? An answer based on quantum measurement theory." *Integrative Physiological and Behavioral Science,* vol. 32, pp. 132–42.

Goswami, U. (2003). *Yoga and Mental Health.* Unpublished manuscript.

Greenwell, B. (1995). *Energies of Transformation.* Saratoga, CA: Shakti River Press.

Grinberg-Zylberbaum, J., Delaflor, M., Attie, L., and Goswami, A. (1994). "Einstein-Podolsky-Rosen paradox in the human brain: The transferred potential." *Physics Essays,* vol. 7, pp. 422-28.

Grof, S. (1992). *The Holotropic Mind.* San Francisco: HarperSan Francisco.

Grossinger, R. (2000). *Planet Medicine.* Berkeley, CA: North Atlantic Books.

Grossman, R. (1985). *The Other Medicines.* Garden City, NY: Doubleday.

Hayflick, L. (1965). "The relative in vitro lifetime of human diploid cell strains." *Experimental Cell Research,* vol. 37, pp. 614–36.

Helmuth, T., Zajonc, A. G., and Walther, H. (1986). In *New Techniques and Ideas in Quantum Measurement Theory,* ed. D. M. Greenberger. New York: New York Academy of Science.

Ho, M. W. (1993). *The Rainbow and the Worm.* Singapore/River Edge, NJ: World Scientific.

Hofstadter, D. R. (1980). *Gödel, Escher, Bach: An Eternal Golden Braid.* New York: Vintage Books.

Holmes, E. (1938). *Science of Mind.* New York: Tarcher/Putnam.

Jahn, R. (1982). "The persistent paradox of psychic phenomena: An engineering perspective." *Proceedings of the IEEE,* vol. 70, pp. 135–70.

Joy, W. B. (1979). *Joy's Way.* Los Angeles, CA: Tarcher.

Jung, C. G. (1971). *The Portable Jung,* ed. J. Campbell. New York: Viking.

Kason, Y. (1994). *Farther Shores.* Toronto: Harper Collins Canada.

Kübler-Ross, E. (ed.) (1975). *Death: The Final Stage of Growth.* Englewood Cliffs, NJ: Prentice-Hall.

Lad, V. (1984). *Ayurveda: The Science of Self-Healing.* Santa Fe, NM: Lotus Press.

Le Fanu, J. (2000). *The Rise and Fall of Modern Medicine.* New York: Carroll & Graf.

Leonard, G. (1990). *The Ultimate Athlete.* Berkeley, CA: North Atlantic Books.

Leviton, R. (2000). *Physician.* Charlottesville, VA: Hampton Roads.

Lewontin, R. (2000). *The Triple Helix.* Cambridge, MA: Harvard University Press.

Libet, B., Wright, E., Feinstein, B., and Pearl, D. (1979). "Subjective referral of the timing of a cognitive sensory experience." *Brain,* vol. 102, p. 193.

Liu, Yen-Chih. (1988). *The Essential Book of Traditional Chinese Medicine.* New York: Columbia University Press.

Locke, S., and Colligan, D. (1986). *The Healer Within.* New York: Dutton.

Lovelock, J. (1982). *Gaia: A New Look at Life on Earth.* New York: Oxford University Press.

Merrell-Wolff, F. (1973). *The Philosophy of Consciousness without an Object.* New York: Julian Press.

———. (1994). *Franklin Merrell-Wolff's Experience and Philosophy.* Albany, New York: SUNY Press.

Mindell, A. (1985). *Working with the Dreaming Body.* New York: Arkana.

Mitchell, M., and Goswami, A. (1992). "Quantum mechanics for observer systems." *Physics Essays,* vol. 5, pp. 525–29.

Moody, R. (1976). *Life After Life.* New York: Bantam.

Moss, R. (1981). *The I That Is We.* Berkeley, CA: Celestial Arts.

———. (1984). *Radical Aliveness.* Berkeley, CA: Celestial Arts.

Motoyama, H. (1981). *Theories of the Chakras.* Wheaton, IL: Theosophical Publishing House.

Moyers, Bill (1993). *Healing and the Mind.* New York: Doubleday.

Nikhilananda, Swami (trans.) (1964). *The Upanishads.* New York: Harper & Row.

Nuland, S. B. (1994). *How We Die.* New York: Knopf.

O'Regan, B. (1987). *Spontaneous Remission: Studies of Self-Healing.* Sausalito, CA: Institute of Noetic Sciences.

————. (1997). "Healing, remission, and miracle cures." In Schlitz, M., and Lewis, N. (ed.), *The Spontaneous Remission Resource Packet.* Sausalito, CA: Institute of Noetic Sciences.

O'Regan, B., and Hirshberg, C. (1993). *Spontaneous Remission: An Annotated Bibliography.* Sausalito, CA: Institute of Noetic Sciences.

Ornish, D. (1992). *Dean Ornish's Program for Reversing Heart Disease.* New York: Ballantine.

Page, C. (1992). *Frontiers of Health.* Saffron Walden, UK: C. W. Daniel.

Pelletier, K. (1981). *Longevity: Fulfilling Our Biological Potential.* New York: Delacorte Press.

————. (1992). *Mind as Healer, Mind as Slayer.* New York: Delta.

————. (2000). *The Best Alternative Medicine.* New York: Simon & Schuster.

Penrose, R. (1989). *The Emperor's New Mind.* New York: Oxford University Press.

Pert, C. (1997). *Molecules of Emotion: Why You Feel the Way You Feel.* New York: Scribner.

Piaget, J. (1977). *The Development of Thought: Equilibration of Cognitive Structures.* New York: Viking.

Posner, M. I., and Raichle, M. E. (1994). *Images of Mind.* New York: Scientific American Library.

Rahe, R. H. (1975). "Epidemiological studies of life change and illness." *International Journal of Psychiatry in Medicine,* vol. 6, pp. 133–46.

Robbins, J. (1996). *Reclaiming Our Health.* Tiburon, CA: H. J. Kramer.

Sabom, M. (1982). *Recollections of Death.* New York: Harper & Row.

Salovey, P., and Mayer, J. D. (1990). "Emotional intelligence." *Imagination, Cognition, and Personality,* vol. 9, pp. 185–211.

Sancier, K. M. (1991). "Medical applications of Qigong and emitted Qi on humans, animals, cell cultures, and plants: Review of selected scientific research." *American Journal of Acupuncture,* vol. 19, pp. 367–77.

Sarno, J. E. (1998). *The Mind-Body Prescription.* New York: Warner Books.

Schlitz, M., and Lewis, N. (1997). *The Spontaneous Remission Resource Packet.* Sausalito, CA: Institute of Noetic Sciences.

Schmidt, H. (1993). "Observation of a psychokinetic effect under highly controlled conditions." *Journal of Parapsychology,* vol. 57, pp. 351–72.

Searle, J. R. (1987). "Minds and brains without programs." In *Mind Waves,* ed. C. Blackmore and S. Greenfield. Oxford, UK: Basil Blackwell.

———. (1994). *The Rediscovery of the Mind.* Cambridge, MA: MIT Press.

Sheehan, M. P. and Atherton, D. J. (1992). "Efficacy of traditional Chinese herbal therapy in adult atopic dermatitis." *Lancet,* vol. 340, pp. 13–17.

Sheldrake, R. (1981). *A New Science of Life.* Los Angeles: Tarcher.

Siegel, B. S. (1990). *Peace, Love, and Healing.* New York: HarperPerennial.

Simonton, O. C., Mathew-Simonton, S., and Creighton, J. J. (1978). *Getting Well Again.* Los Angeles: Tarcher.

Sperry, R. (1983). *Science and Moral Priority.* New York: Columbia University Press.

Squires, E. J. (1987). "A viewer's interpretation of quantum mechanics." *European Journal of Physics,* vol. 8, pp. 171–174.

Stapp, H. P. (1994). *A Report on the Gaudiya Vaishnave Vedanta*. San Francisco, CA: Bhaktivedanta Institute.

———. (1995). "The hard problem: A quantum approach." *Journal of Consciousness Studies,* vol. 3, pp. 194–200.

Stevenson, I. (1987). *Children Who Remember Previous Lives: A Question of Reincarnation*. Charlottesville: University Press of Virginia.

Svoboda, R., and Lade, A. (1995). *Tao and Dharma: A Comparison of Ayurveda and Chinese Medicine*. Twin Lakes, WI: Lotus Press.

Taimni, I. K.: (1961). *The Science of Yoga*. Wheaton, IL: Theosophical Publishing House.

Teilhard de Chardin, P. (trans. Bernard Wall) (1959). *The Phenomenon of Man*. New York: Harper.

Tipler, F. (1994). *The Physics of Immortality: Modern Cosmology, God, and the Resurrection of the Dead*. New York: Doubleday.

Ullman, D. (1988). *Homeopathy: Medicine for the 21st Century*. Berkeley, CA: North Atlantic Books.

van Lomel, P., van Wees, R., Meyers, V., Elfferich, I. (2001). "Near-death experiences in survivors of cardiac arrest." *Lancet,* vol. 358, pp. 2039–45.

Vithoulkas, G. (1980). *The Science of Homeopathy*. New York: Grove Press.

von Neumann, J. (1955). *The Mathematical Foundations of Quantum Mechanics*. Princeton: Princeton University Press.

Waddington, C. (1957). *The Strategy of the Genes*. London: Allen & Unwin.

Wallas, G. (1926). *The Art of Thought*. New York: Harcourt, Brace & Co.

Weil, A. (1983). *Health and Healing*. Boston: Houghton Mifflin.

———. (1995). *Spontaneous Healing*. New York: Knopf.

Wilber, K. (1993). "The great chain of being." *Journal of Humanistic Psychology,* vol. 33, pp. 52–55.

Williams, R. (1989). *The Trusting Heart: Great News about Type A Behavior.* New York: Times Books.

Wolf, F. A. (1986). *The Body Quantum.* New York: Macmillan.

———. (1988). *Parallel Universes.* New York: Simon & Schuster.

———. (1996). *The Spiritual Universe.* New York: Simon & Schuster.

Index

About the Author

Amit Goswami is a theoretical nuclear physicist. Goswami received his PhD in physics from the University of Calcutta in 1964 and moved to the United States early in his career. He taught physics for 32 years as a member of The University of Oregon Institute for Theoretical Physics. Starting at age 38, his research interests shifted to quantum cosmology, quantum measurement theory, and applications of quantum mechanics to the mind-body problem. These days he is probably best known as one of the interviewed scientists featured in the 2004 film *What the Bleep Do We Know!?* He is also featured in the recent documentary *Dalai Lama Renaissance* and is the subject of the 2009 documentary *The Quantum Activist.*

Fully retired as a faculty member since 2003, Dr. Goswami now speaks nationally and internationally. He is also a member of the advisory board of the Institute of Noetic Sciences, where he was a senior scholar in residence from 1998 to 2000.

Find out more about Dr. Goswami at *amitgoswami.org*

Hampton Roads Publishing Company

. . . for the evolving human spirit

Hampton Roads Publishing Company
publishes books on a variety of subjects,
including spirituality, health,
and other related topics.

For a copy of our latest trade catalog,
call 978-465-0504, or visit our website at
www.hrpub.com